WATER RESOURCE PLANNING, DEVELOPMENT AND MANAGEMENT

INDIAN IRRIGATION PROJECTS

MAINTENANCE AND SUSTAINABILITY ISSUES

WATER RESOURCE PLANNING, DEVELOPMENT AND MANAGEMENT

WATER RESOURCE PLANNING, DEVELOPMENT AND MANAGEMENT

INDIAN IRRIGATION PROJECTS

MAINTENANCE AND SUSTAINABILITY ISSUES

CHRISTOPHER FLORES
EDITOR

nova publishers
New York

NOTICE TO THE READER

The Publisher has taken reasonable care in the preparation of this book, but makes no expressed or implied warranty of any kind and assumes no responsibility for any errors or omissions. No liability is assumed for incidental or consequential damages in connection with or arising out of information contained in this book. The Publisher shall not be liable for any special, consequential, or exemplary damages resulting, in whole or in part, from the readers' use of, or reliance upon, this material. Any parts of this book based on government reports are so indicated and copyright is claimed for those parts to the extent applicable to compilations of such works.

Independent verification should be sought for any data, advice or recommendations contained in this book. In addition, no responsibility is assumed by the publisher for any injury and/or damage to persons or property arising from any methods, products, instructions, ideas or otherwise contained in this publication.

This publication is designed to provide accurate and authoritative information with regard to the subject matter covered herein. It is sold with the clear understanding that the Publisher is not engaged in rendering legal or any other professional services. If legal or any other expert assistance is required, the services of a competent person should be sought. FROM A DECLARATION OF PARTICIPANTS JOINTLY ADOPTED BY A COMMITTEE OF THE AMERICAN BAR ASSOCIATION AND A COMMITTEE OF PUBLISHERS.

Additional color graphics may be available in the e-book version of this book.

Library of Congress Cataloging-in-Publication Data

ISBN: 978-1-63483-261-8

Published by Nova Science Publishers, Inc. † New York

CONTENTS

PREFACE

Over 100 irrigation projects and systems can be found on Indian reservations primarily across the western United States. The scarcity of water in much of the western United States makes irrigation critical to agricultural activities. The Department of the Interior's Bureau of Indian Affairs (BIA) manages 16 irrigation projects on Indian reservations in the western United States. This book focuses on BIA's estimated deferred maintenance cost for its irrigation projects; shortcomings that the Government Accountability Office (GAO) identified in BIA's management of its irrigation projects; and issues GAO identified that needed to be addressed to determine the long-term direction of BIA's irrigation program.

In: Indian Irrigation Projects
Editor: Christopher Flores
ISBN: 978-1-63483-261-8

Chapter 1

INDIAN IRRIGATION PROJECTS: DEFERRED MAINTENANCE AND FINANCIAL SUSTAINABILITY ISSUES REMAIN UNRESOLVED*

Anne-Marie Fennell

WHY GAO DID THIS STUDY

Over 100 irrigation projects and systems can be found on Indian reservations primarily across the western United States. The scarcity of water in much of the western United States makes irrigation critical to agricultural activities. In February 2006, GAO reported on 16 irrigation projects where BIA charged water users for the projects' operation and maintenance (GAO-06-314). These projects, which were generally constructed in the late 1800s and early 1900s, included water storage facilities and delivery structures for agricultural purposes.

This testimony is based on GAO's February 2006 report and updated information on BIA's fiscal year 2014 estimate of deferred maintenance and

* This is an edited, reformatted and augmented version of a report, GAO-15-453T, issued by the U.S. Government Accountability Office, March 4, 2015. This report was originally presented as testimony before the Senate Committee on Indian Affairs. Anne-Marie Fennell is Director, Natural Resources and Environment at the GAO.

actions BIA has taken to address GAO's three recommendations. The testimony focuses on (1) BIA's estimated deferred maintenance cost for its irrigation projects, (2) shortcomings that GAO identified in BIA's management of its irrigation projects, and (3) issues GAO identified that needed to be addressed to determine the long-term direction of BIA's irrigation program.

GAO is not making any new recommendations in this testimony.

WHAT GAO FOUND

The Department of the Interior's Bureau of Indian Affairs (BIA) estimated the cost for deferred maintenance for the 16 irrigation projects covered in GAO's February 2006 report at about $850 million for fiscal year 2005. To further refine the estimate, BIA planned to hire engineering and irrigation experts to conduct thorough condition assessments of the irrigation projects to correctly identify deferred maintenance needs and costs. While the irrigation projects included in the estimate have changed somewhat in the 9 years since GAO's report, BIA's fiscal year 2014 cost estimate for deferred maintenance for its irrigation projects is about $570 million.

In its February 2006 report, GAO found BIA's management of some of its irrigation projects had serious shortcomings that undermined effective decision making about project operations and maintenance. First, under BIA's organizational structure, officials with the authority to oversee irrigation project managers generally lacked the technical expertise needed to do so effectively, while the staff that had the expertise lacked the necessary authority to oversee project managers' decision making. Second, BIA had not consistently provided project stakeholders, such as water users, with the necessary information or opportunities to participate in project decision making, contrary to federal regulations that required BIA to consult with project stakeholders in setting project priorities. BIA has implemented GAO's two recommendations related to these management shortcomings.

In its February 2006 report, GAO found that the long-term direction of BIA's irrigation program depended on the resolution of several larger issues.

- *Financial sustainability.* BIA did not know to what extent its irrigation projects were capable of financially sustaining themselves, hindering its ability to address long-standing concerns regarding inadequate funding.

- *Funding for deferred maintenance.* BIA did not have a plan for how to obtain funding to fix deferred maintenance items—a significant challenge in times of tight budgets and competing priorities.
- *Alternative project managers.* Given BIA's many responsibilities in support of Indian communities, it might be more appropriate for other entities, such as tribes or water users, to manage some or all of the irrigation projects.

To obtain information on the long-term financial sustainability of each of the projects, GAO recommended that BIA conduct studies to determine how much it would cost to financially sustain each project and the extent to which water users on each project have the ability to pay these costs. Subsequently, in June 2008, the Department of the Interior stated in a memorandum that it did not have sufficient funding to perform these studies—and did not expect to have such funding in the foreseeable future. Since GAO's February 2006 report, BIA irrigation projects continue to face hundreds of millions of dollars of deferred maintenance needs, and financial sustainability issues also remain unresolved.

* * *

Chairman Barrasso, Vice Chairman Tester, and Members of the Committee:

I am pleased to be here today to participate in your hearing on S. 438— a bill to provide for the repair, replacement, and maintenance of certain Indian irrigation projects. There are over 100 irrigation projects and systems on Indian reservations primarily across the western United States. As you know, the scarcity of water in much of the western United States makes irrigation critical to the continued success of agricultural activities. In February 2006, we reported on 16 Indian irrigation projects where water users were charged for project operations and maintenance by the Department of the Interior's (Interior) Bureau of Indian Affairs (BIA), which is responsible for providing social and economic services to Indians as well as managing land and natural resources held in trust by the United States for Indians.[1]

Generally initiated in the late 1800s and early 1900s by Interior as part of the federal government's Indian assimilation policy, BIA's irrigation program was designed to foster agricultural opportunities and provide economic benefits to Indian communities. The 16 irrigation projects include water storage facilities and delivery structures for agricultural purposes. Over time, non-Indians began buying or leasing the land served by the projects for agricultural purposes, and project stakeholders evolved from Indian water users and the tribes within the reservations to include non-Indian water users as well. Many of the water users today are non-Indian.

Reports by Interior's Inspector General on BIA's irrigation projects have documented that the annual operations and maintenance fees BIA has charged water users have historically been set too low to cover the full cost of running the projects.[2] In addition, problems have been reported with collecting the fees that have been assessed. Because of insufficient funding, project maintenance has been consistently postponed, resulting in an extensive and costly list of deferred maintenance items. This deferred maintenance ranges from repairing or replacing dilapidated irrigation structures to clearing weeds from irrigation ditches. In addition to the deferred maintenance, water users had expressed concern that BIA had been unresponsive in addressing the projects' ongoing operations and maintenance needs.

My testimony today will summarize the findings of our February 2006 report, along with some recent updates. Specifically, I will discuss

1. BIA's estimated deferred maintenance costs for its irrigation projects;
2. shortcomings that we identified in BIA's management of its irrigation projects; and (3) issues we identified that needed to be addressed to determine the long-term direction of BIA's irrigation program. In addition, I will provide information on actions, where applicable, that BIA has taken to address the three recommendations in our February 2006 report.

For our February 2006 report, we collected documentation from BIA headquarters irrigation officials on the 16 irrigation projects, and we visited and collected information from each of BIA's four regional offices that oversee the 16 irrigation projects. We also visited 9 of the 16 projects, where we collected project-specific information from BIA officials and project stakeholders.[3] To examine estimated deferred maintenance costs, we reviewed BIA's lists of deferred maintenance items and cost estimates, as well as the methodology BIA used to develop these lists and estimates. To determine whether management shortcomings existed, we reviewed relevant federal regulations and agency guidance and we analyzed BIA-wide and project-specific management protocols and systems for the 9 projects we visited. Finally, to determine any issues that needed to be addressed to determine the long-term direction of the projects, we reviewed prior studies on BIA's irrigation program, and we discussed the long-term direction of the program with BIA irrigation officials and project stakeholders. A detailed description of our scope and methodology is presented in appendix I of the February 2006 report.

For comparison purposes and to show changes that BIA has made to its estimate of deferred maintenance costs since our February 2006 report, we collected the most recent estimate of deferred maintenance costs from BIA—data for fiscal year 2014 as of September 30, 2014. We did not assess the reliability of the fiscal year 2014 estimate. We also present information on the status of the three recommendations from our report. The report upon which this testimony statement is based was conducted in accordance with generally accepted government auditing standards. Those standards require that we plan and perform the audit to obtain sufficient, appropriate evidence to provide a reasonable basis for our findings and conclusions based on our audit objectives. We believe that the evidence obtained for our report provided a reasonable basis for our findings and conclusions based on our audit objectives.

BACKGROUND

BIA's irrigation program was initiated in the late 1800s, as part of the federal government's Indian assimilation policy, and it was originally designed to provide economic development opportunities for Indians through agriculture. The Act of July 4, 1884, provided the Secretary of the Interior $50,000 for the general development of irrigation on Indian lands.[4] Over the years, Congress continued to pass additional legislation authorizing and funding irrigation facilities on Indian lands.

BIA's irrigation program includes over 100 "irrigation systems" and "irrigation projects" that irrigate over 750,000 acres primarily across the West. BIA's irrigation systems are nonrevenue-generating facilities that are primarily used for subsistence gardening and are operated and maintained through a collaborative effort, which generally involves other BIA programs, tribes, and water users. In contrast, BIA's 16 irrigation projects that we reported on in our February 2006 report charged their water users an annual operations and maintenance fee to fund the cost of operating and maintaining the project. Most of BIA's irrigation projects have been considered self-supporting through these operations and maintenance fees. The 16 irrigation projects are located on Indian reservations across the agency's Rocky Mountain, Northwest, Southwest, and Western regions (see fig. 1).

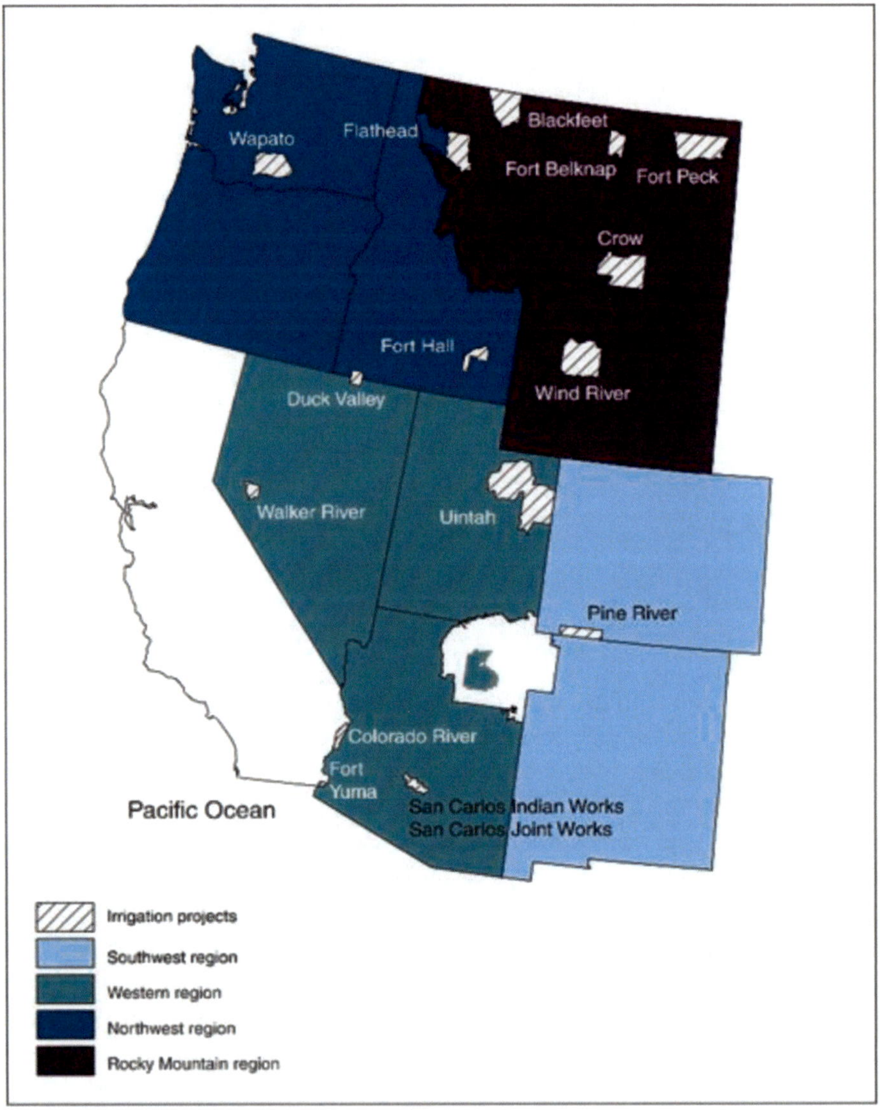

Source: Bureau of Indian Affairs; GAO. | GAO-15-153T.

Figure 1. Location of the 16 Bureau of Indian Affairs Irrigation Projects Reported on by GAO in February 2006.

BIA's management of its irrigation projects was decentralized, with regional and local BIA offices responsible for day-to-day operations and maintenance. Fourteen projects included in our February 2006 report were overseen by local BIA agency superintendents, and the 2 largest projects were overseen directly by regional directors. The local agency superintendents that oversaw these projects reported to their respective regional director. BIA's irrigation and engineering experts, who provide technical assistance to the projects, were located in each region, as well as in BIA's central office located in Washington, D.C., and other BIA locations in the western United States. The regional irrigation staff and central irrigation office staff did not have line authority over the projects.

The irrigation facilities constructed by BIA include a range of structures for storing and delivering water for agricultural purposes. Figure 2 highlights an example of the key structural features found on BIA's irrigation projects.

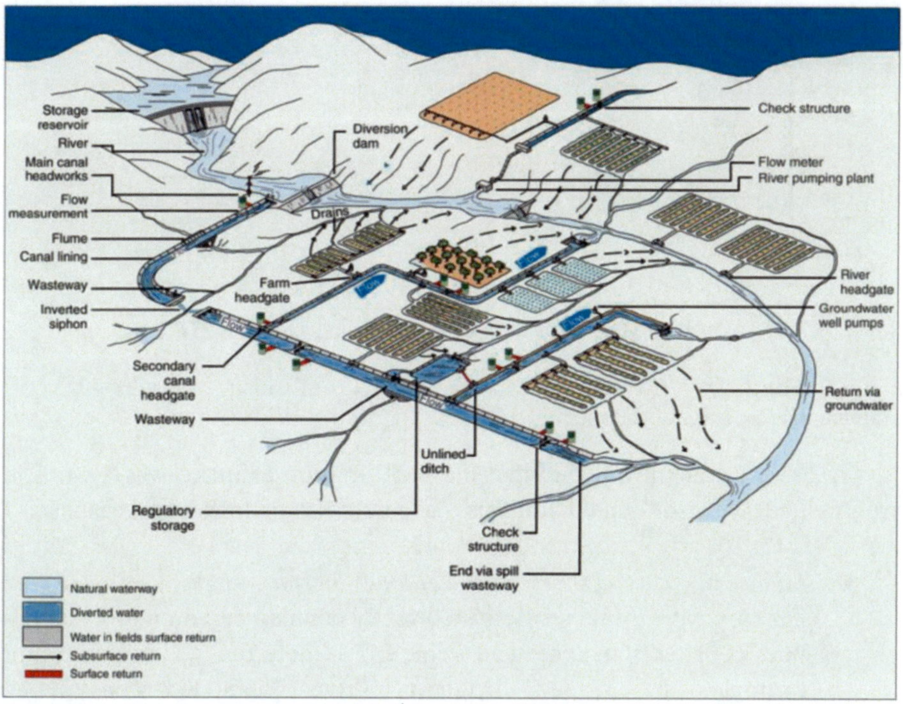

Source: Bureau of Reclamation; GAO. | GAO-15-453T.

Figure 2. Example of an Irrigation Project Operated by the Bureau of Indian Affairs.

BIA Estimated the Cost of Deferred Maintenance at about $850 Million in 2005, but the Estimate Has Since Been Refined to about $570 Million

In our February 2006 report,[5] we found that BIA had estimated the cost for deferred maintenance at the 16 irrigation projects at about $850 million for fiscal year 2005. See figure 3 for a breakdown of the cost estimate by project at that time.

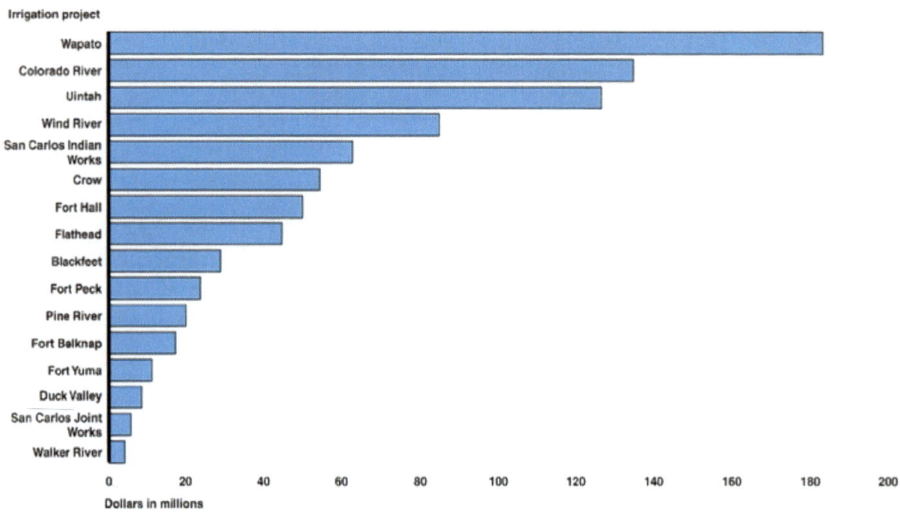

Source: GAO analysis of Bureau of Indian Affairs data. | gao-15-453T.

Figure 3. Fiscal Year 2005 Bureau of Indian Affairs Cost Estimate of Deferred Maintenance by Irrigation Project.

In 2006, we acknowledged that the $850 million estimate was a work in progress, but we also found that it was inaccurate for the following reasons:

- *Some projects incorrectly counted certain items as deferred maintenance***.** Some projects incorrectly counted certain items, such as new construction items and vehicles, as deferred maintenance. For example, the Wapato Irrigation Project included constructing reservoirs and the San Carlos Indian Works Irrigation Project included building a new office. In addition, some projects included the cost of repairing vehicles or buying new ones in their deferred maintenance estimates, despite BIA guidance at the time of our report that such items were not deferred maintenance. For example, the Wind River Irrigation Project included an excavator vehicle, and the Crow Irrigation Project included dump trucks.

- *Some projects provided BIA with incomplete information.* According to BIA officials, some projects did not do thorough assessments of their deferred maintenance needs, and some may not have included legitimate deferred maintenance items, such as resloping canal banks that have eroded by crossing cattle or overgrown vegetation. Moreover, neither the Walker River Irrigation Project nor the Uintah Irrigation Project provided information detailing their deferred maintenance costs at the time of our report.
- *BIA made errors when compiling the total deferred maintenance cost estimates.* For example, BIA inadvertently double-counted the estimate provided by the Colorado River Irrigation Project when compiling the overall cost estimate, according to BIA officials. Additionally, BIA officials erroneously estimated costs for all structures, such as flumes and check gates, based on the full replacement values even when items were in good or fair condition and needed only repairs.

In 2006, we concluded that while the inclusion of incorrect items and calculation errors likely contributed to the overestimation of BIA's total deferred maintenance costs, the incomplete information provided to BIA by some projects may have contributed to the underestimation of the total costs.

As we reported in 2006, to further refine its cost estimate and to develop more comprehensive deferred maintenance lists, BIA planned to hire experts in engineering and irrigation to periodically conduct thorough condition assessments of all 16 irrigation projects to identify deferred maintenance needs and costs. According to BIA officials, these thorough condition assessments were expected to more accurately reflect each project's actual deferred maintenance, in part because experts in engineering and irrigation who can differentiate between structural and cosmetic problems were to conduct them. These assessments were also to help BIA prioritize the allocation of potential funds to complete deferred maintenance items because they would assign a prioritization rating to each deferred maintenance item based on the estimated repair or replacement cost, as well as the overall importance to the project. The first such assessment was completed in July 2005, and BIA planned to reassess the condition of each project at least once every 5 years, with the first round of such condition assessments to be completed by the end of 2010.

While the irrigation projects included in BIA's estimate of deferred maintenance costs have changed somewhat since our report, the most recent deferred maintenance cost estimate for fiscal year 2014 was just under $570 million (see table 1).[6] Several reasons may have contributed to the lower estimate including more thorough condition assessments and maintenance work performed since our report. However, we did not assess the reliability of the fiscal year 2014 estimate.[7] The new estimate is presented for comparison purposes to demonstrate changes that BIA made to the earlier fiscal year 2005 estimate that we raised concerns about in our February 2006 report. Table 1 also shows that most of the condition assessments are now more than 5 years old, and they were not all completed by 2010. Condition assessments for a few projects are still ongoing.

Table 1. Bureau of Indian Affair's Refined Estimate of Deferred Maintenance Costs, Fiscal Year 2014

Dollars in millions			
Irrigation project	*Deferred maintenance for fiscal year 2014*	*Condition assessment completion date*	*Replacement value*
Irrigation projects in GAO's February 2006 report, GAO-06-314			
Wapato	$138	2014	$1,371
Flathead	82	2008	237
Fort Hall	30	2009	128
Wind River	30	2008	93
Blackfeet	26	2007	50
Uintah	25	2009	593
Colorado River	17	2011	332
Crow	17	2007	59
Walker River	17	2008	44
San Carlos[a]	16	Ongoing	201
Fort Peck	13	2007	33
Pine River	11	2008	39
Duck Valley	8	2009	48
Fort Belknap	8	2007	19
Fort Yuma	[b]	[b]	[b]
Subtotal	*$438*		*$3,247*
Other irrigation real property assets that the Bureau of Indian Affairs counts in its estimate			
Navajo	$125	Ongoing	$1,039
Pyramid Lake	5	Ongoing	49

Dollars in millions			
Irrigation project	*Deferred maintenance for fiscal year 2014*	*Condition assessment completion date*	*Replacement value*
Subtotal	*$130*		*$1,088*
Total	*$568*		*$4,335*

Source: Bureau of Indian Affairs. | GAO-15-453T.

Notes: We did not assess the reliability of BIA's new estimates, they are presented simply for comparison purposes to demonstrate changes that BIA has made since its fiscal year 2005 estimate.

Some water rights settlements between Indian tribes and the federal government established funds or accounts for irrigation project maintenance, which may be used to fund some or all of the deferred maintenance at those projects. For example, water rights settlements between the federal government and the Crow Tribe and the Shoshone-Paiute Tribes of the Duck Valley Reservation established funds or accounts for maintenance of the Crow and Duck Valley Irrigation Projects, respectively.

[a] In our February 2006 report, we listed the San Carlos Joint Works Irrigation Project and the San Carlos Indian Works Irrigation Project separately. Since February 2006 report, the management of these projects has been reorganized, and the Bureau of Indian Affairs now tracks them together as one project.

[b] Fort Yuma is the responsibility of the Bureau of Reclamation.

BIA ADDRESSED THE MANAGEMENT SHORTCOMINGS IDENTIFIED IN OUR FEBRUARY 2006 REPORT

In our February 2006 report, we found that BIA's management of some of its irrigation projects had serious shortcomings that undermined effective decision making about project operations and maintenance. First, under BIA's organizational structure, in many cases, officials with the authority to oversee project managers' decision making lacked the technical expertise needed to do so effectively, while the staff who had the expertise lacked the necessary authority to oversee project managers' decision making. The BIA regional directors and local agency superintendents and deputy superintendents that provided oversight on projects did not generally have engineering or irrigation expertise and relied heavily on the project managers to run the projects. Of the nine projects that we visited for our February 2006 report, only two had managers at the regional or agency levels who were experts in irrigation or engineering. We found that such an organizational structure and reliance on the project managers breaks down when the person managing the project lacks the expertise required for the position—that is, in cases in which BIA has had difficulty filling project manager vacancies and has, as a result, hired less qualified people. For example, at the Crow project in 2002, a project manager with insufficient expertise decided to repair a minor leak in a key water delivery structure by dismantling it and replacing it with a different type of structure. The new structure was subsequently deemed inadequate by BIA's irrigation experts, and the required reconstruction delayed water delivery by about a month. Furthermore, we found that the BIA staff with the necessary expertise— regional irrigation engineers and central irrigation office staff—had no authority over the 16 projects. Consequently, key technical decisions about project operations and maintenance, such as when or how to repair critical water delivery infrastructure, did not necessarily get the technical oversight or scrutiny needed.

To address this shortcoming, in our February 2006 report, we recommended that BIA provide the necessary level of technical support for project managers who have less than the desired level of engineering qualifications by putting these projects under the direct supervision of regional or central irrigation office staff or by implementing more stringent protocols for engineer review and approval of actions taken at the projects. In response to our recommendation, in February 2007, the Director of BIA issued a technical review and assistance policy directive to the relevant BIA regional directors to ensure that adequate review and assistance is given to BIA irrigation project managers. The policy provided for strict protocols for engineer review and approval of actions taken at the projects by those with the necessary engineering expertise. It also outlined specific responsibilities for irrigation project managers, as well as other key irrigation staff. In addition, BIA has made other organizational line authority changes to address this recommendation.[8]

Second, in our February 2006 report, we found that BIA did not consistently provide information and opportunities for stakeholders to participate in setting project priorities. Federal regulations required BIA to consult with project stakeholders—such as, tribal council representatives, as well as Indian and non-Indian water users—in setting project priorities but BIA did not consistently do so. [9] For example, we reported that the Wapato Irrigation Project had shared little information on its spending with stakeholders, and the Pine River Irrigation Project did not meet with its nontribal stakeholders, limiting stakeholders' ability to have an impact on project decisions and BIA's ability to benefit from their input.

To address the second shortcoming, in our February 2006 report, we recommended that BIA require, at a minimum, that irrigation project management meet twice annually with all project stakeholders—once at the end of a season and once before the next season—to provide information on project operations, including budget plans and actual annual expenditures, and to obtain feedback and input. In response to our recommendation, in July 2006, the Acting Director of BIA directed each of the four BIA regional directors responsible for the 16 irrigation projects to personally ensure that irrigation staff meet twice annually, at a minimum, with water users and other stakeholders—once at the end of the season and once before the next season. For projects that operate year-round, the project managers in consultation with project water users were to determine mutually acceptable times for holding these two annual meetings. At these meetings, BIA's irrigation project managers and irrigation staff were directed to provide information on project operations, including budget plans and actual annual expenditures, and obtain feedback and input. This policy change was published in the *Federal Register* in April 2007.[10] In addition, BIA irrigation project managers were directed to submit documentation of the meetings to BIA headquarters irrigation staff.

In our February 2006 report, we found that the long-term direction of BIA's irrigation program depended on the resolution of the following larger issues:

- Financial sustainability. Of the most importance, BIA did not know to what extent its irrigation projects were capable of financially sustaining themselves, which hindered the agency's ability to address long-standing concerns regarding inadequate funding. Despite this lack of information on the overall financial situation for each of the projects, in the early 1960s, BIA classified more than half of the 16 projects that we reported on as fully self-supporting on the basis of annual operations and maintenance fees they collected from water users. These self-supporting projects did not receive any ongoing appropriated funds. These projects were subject to full cost recovery despite the absence of financial information to demonstrate that the water users could sustain this financial burden. The heavy reliance on water users to sustain these projects had created ongoing tension between the water users and BIA. Some water users had complained to BIA that they could not afford the operations and maintenance fees, and they had pressured BIA to keep the fees as low as possible. Without definitive information on the financial situation of each project, we concluded that BIA could not determine what portion of project operations and maintenance costs can be reasonably borne by the water users and to what extent alternative sources of financing, such as congressional appropriations, should be pursued.

- Funding for deferred maintenance. The future of BIA's irrigation program also depended on the resolution of how the deferred maintenance will be funded. BIA did not have a plan for how it would obtain funding to fix the deferred maintenance items. Regardless of the precise cost estimate for total deferred maintenance, we concluded that funding deferred maintenance costs in the hundreds of millions of dollars will be a significant challenge in times of tight budgets and competing priorities.
- Alternative project managers. Given that BIA must balance irrigation management with its many other missions in support of Indian communities, such as providing education and law enforcement, we reported that there were inherent limits on the resources and knowledge that BIA was able to devote to any one program. As a result of these limitations and competing demands, officials told us at the time of our report that irrigation management is not a priority for BIA. In our February 2006 report, we found that it may be beneficial to consider whether others for whom irrigation is more of a priority or an area of expertise, including other federal agencies, Indian tribes, and water users, could better manage some of the projects. We concluded that successful management of the projects by others, however, would depend on the characteristics of each project and its stakeholders. For example, turning over projects to tribes may be an option for projects where most of the water users are Indian, whereas turning over projects to water users may be an option for projects where water users share similar interests and have a desire to organize into an irrigation district or association.

To obtain information on the long-term financial sustainability of each of the projects, we recommended in our February 2006 report that BIA conduct studies to determine both how much it would cost to financially sustain each project, and the extent to which water users on each project have the ability to pay these costs.[11] We stated that this information would be useful to congressional decision makers and other interested parties in debating the long-term direction of BIA's irrigation program. However, to date, BIA has not implemented this recommendation. In June 2008, the Department of the

Interior provided us with a memorandum that stated, while the department agreed that studies to assess the financial sustainability of the irrigation projects would be valuable, it did not have sufficient funding to perform these studies—and does not expect to have such funding in the foreseeable future.

In conclusion, BIA irrigation projects continue to face hundreds of millions of dollars of deferred maintenance needs. The Senate bill, S. 438, if enacted, could help address these needs and potentially some of the other larger issues that we reported on in our February 2006 report. By establishing an Indian Irrigation Fund for fiscal years 2015 through 2036, this bill, if enacted, would help provide needed resources to carry out maintenance, repair, and replacement activities for certain Indian irrigation projects and funds to conduct a study of BIA's Indian irrigation program and project management.

Chairman Barrasso, Vice Chairman Tester, and Members of the Committee, this completes my prepared statement. I would be pleased to answer any questions that you may have at this time.

End Notes

[1] GAO, Indian Irrigation Projects: Numerous Issues Need to Be Addressed to Improve Project Management and Financial Sustainability, GAO-06-314 (Washington, D.C.

[2] Department of the Interior, Office of the Inspector General, Indian Irrigation Projects, Bureau of Indian Affairs, 96-I-641 (Washington D.C.: March 1996); Department of the Interior, Office of the Inspector General, Operations and Maintenance Assessments of Indian Irrigation Projects, Bureau of Indian Affairs, W-IA-BIA-12-86 (Washington D.C.: February 1988).

[3] We selected these projects based on a combination of factors aimed at maximizing our total coverage (over 50 percent of the projects), visiting at least one project in each of the regions where irrigation projects are located, visiting the project with the highest deferred maintenance cost estimate in each region using BIA's fiscal year 2004 data, and visiting what BIA considered to be the three best projects and the five worst projects. Specifically, we visited: (1) the Blackfeet Irrigation Project, (2) the Colorado River Irrigation Project, (3) the Crow Irrigation Project, (4) the Fort Belknap Irrigation Project, (5) the Pine River Irrigation Project, (6) the San Carlos Indian Works Irrigation Project, (7) the San Carlos Joint Works Irrigation Project, (8) the Wapato Irrigation Project, and (9) the Wind River Irrigation Project.

[4] Act of July 4, 1884, 23 Stat. 76, 94 (1884).

[5] GAO-06-314.

[6] The fiscal year 2014 and fiscal year 2005 estimates, either by project or in total, cannot be directly compared without adjusting for inflation. In addition, since the number of projects included in each year's total estimate varies, any comparison of the total estimates would not be meaningful.

[7] Specifically, we did not determine the extent to which BIA addressed the inaccuracies that we found in its fiscal year 2005 estimate or whether any such inaccuracies still remain.

[8] For example, according to BIA, the Rocky Mountain Region realigned the organizational structure for its five irrigation projects. The five Irrigation Project Managers now report directly to the Regional Water Resources Branch Chief, an engineering position.

[9] 25 C.F.R. § 171.1(c) (2005). This regulation was amended in 2008 to require BIA to cooperate and consult with all interested parties, especially persons or entities to which it provides irrigation service and receives uses of BIA irrigation facilities, such as irrigators and landowners. 25 C.F.R. § 171.110(b) (2015).

[10] 72 Fed. Reg. 19950 (Apr. 20, 2007).

[11] GAO-06-314.

In: Indian Irrigation Projects
Editor: Christopher Flores
ISBN: 978-1-63483-261-8

Chapter 2

STATEMENT OF KEVIN WASHBURN, ASSISTANT SECRETARY INDIAN AFFAIRS, U.S. DEPARTMENT OF THE INTERIOR. LEGISLATIVE HEARING ON "S. 438, THE IRRIGATION REHABILITATION AND RENOVATION FOR INDIAN TRIBAL GOVERNMENTS AND THEIR ECONOMIES ACT, THE IRRIGATE ACT"*

GoodafternoonChairmanBarrasso,ViceChairmanTester,andmembersofthe Committee.My name is Kevin Washburn and I am the Assistant Secretary for Indian Affairs at the Department of the Interior (Department). Thank you for inviting the Department to provide testimony on S. 438, the Irrigation Rehabilitation and Renovation for Indian Tribal Governments and Their Economies Act, a bill to provide for the repair, replacement, and maintenance of certain Indian irrigation projects. We appreciate the Committee's continued leadership on the daunting challenge the Department faces on addressing the deferred maintenanceat the 17 Bureau of Indian Affairs' (BIA)Irrigation projects.

* This is an edited, reformatted and augmented version of a statement presented March 4, 2015 before the Senate Committee on Indian Affairs.

Larry Roberts, Principal Deputy Assistant Secretary for Indian Affairs, for the Department, testified before this Committee last September and provided an overview of the Irrigation Projects in Indian Country, along with the BIA's Irrigation Program's accomplishments, Irrigation project condition assessments, and the deferred maintenance estimates for our 17 BIA Irrigation Projects. S. 438 seeks to address the deferred maintenance for the BIA's Irrigation Projects by identifying the eligible projects, establishing a priority for allocating $35 million per year to our BIA Irrigation projects for twenty-two years to carry out the maintenance, repair and replacement activities at the irrigation projects. S. 438 also provides for establishing programmatic goals and conducting a study aimed at improving program and project management, and performance of BIA irrigation projects with a report to be delivered to Congress within 2 years. The Department supports the goals of working with tribes to address the maintenance of irrigation projects, and we look forward to working with you to address the best means of doing so given current budget constraints and the ability of irrigation projects to financially sustain themselves in the long run.

BACKGROUND

The Federal government has been involved with Indian irrigation since the Colorado River Indian Irrigation Project was authorized in 1867. In the early 1900's, Congress began authorizing funding for the construction of numerous Indian irrigation projects in the western United States. At that time, the Indian Irrigation Service led construction and early administration of the projects. In the late 1930's and through the 1940's, as construction activities wrapped up on most projects, the Indian Irrigation Service ceased to exist and operation and maintenance (O&M) was transferred to the BIA, where it continues today. Many of these programs began at a time when Federal policies were far different. These irrigation projects remain very important today to the communities they serve. The BIA irrigation program is responsible for oversight and administration of fifteen revenue-generatingIndian irrigation projects that provide service and water delivery to over 25,000 customers and 750,000 acres of land in Indian Country. The asset inventory and program responsibilities also include BIA-owned facilities at non-revenue generating irrigation projects, including the Navajo Indian Irrigation Project in New Mexico and Pyramid Lake Irrigation Project in northern Nevada.At these facilities the BIA does not assess O&M charges to irrigators; those charges are instead paid through appropriations or other means.

S. 438

S. 438 would create an "Indian Irrigation Fund" (IIF) in the Department of the Treasury from the reclamation fundthat was established in the Act of June 17, 1902 (32 Stat. 388, chapter 1093). The IIF would be funded at $35 million per year for 22 years for a total investment of $770 million. The legislation also caps the year to year spending at $35million, but includes amounts of interest earned on investments from the IIF, if applicable, "to carry out maintenance, repair and replacement activities" for one or more of the eligible Indian irrigations projects identified in Sec. 202 of the S. 438.

Section 104 requires the Secretary of the Interior to invest portions of the IFF that in the judgment of the Secretary are not required to meet current withdrawals. We recommend that the U.S. Treasury be designated as the federal agency responsible for investing IFF assets. We need to consult with the Department of the Treasury in more detail about these provisions.

Section 201(b) of S. 438 describes funding "to carry out maintenance, repair and replacement activities….". The term "maintenance" is used and further states "including any structures, facilities, equipment, or vehicles used in connection with the operation of those projects." We interpret this language asauthorizing fundingfor the purchase of heavy equipment to address some deferred maintenance items, conduct routine and preventative maintenance activities, and also to purchase vehicles to support water delivery/operation activities. If this is the case we request adding the term "personnel" to the list of items in parenthesis in Section 201(b) in order to clarify that hiring personnel is allowable in supporting O&M of the eligible projects.

Section 202 of S. 438 defines the eligible projects for the IIF. The following is a list of the 17 Irrigation Projects that meet the criteria listed in Section 202.

- Blackfeet Indian Irrigation Project (MT)
- Colorado River Indian Irrigation Project (AZ/CA)
- Crow Indian Irrigation Project (MT)
- Duck Valley Indian Irrigation Project (NV)
- Flathead Indian Irrigation Project (MT)

- Fort Belknap Indian Irrigation Project (MT)
- Fort Hall Indian Irrigation Project (ID)
- Fort Peck Indian Irrigation Project (MT)
- Navajo Indian Irrigation Project (NM)
- Pine River Indian Irrigation Project (CO)
- Pyramid Lake Indian Irrigation Project (NV)
- San Carlos Indian Irrigation Project – Indian Works (AZ)
- San Carlos Indian Irrigation Project – Joint Works (AZ)
- Uintah Indian Irrigation Project (UT)
- Walker River Indian Irrigation Project (NV)
- Wapato Indian Irrigation Project (WA)
- Wind River Indian Irrigation Project (WY)

We recommend amending Section 202(2), as follows: "are managed and operated by the Bureau of Indian Affairs (including projects managed, operated and/or maintained under contracts or compacts pursuant to the Indian Self-Determination and Education Assistance Act (25 U.S.C. 450 et seq.) *or other agreements with water users, water user groups and/or water user associations*; and…." This is to ensure that the irrigation projects with 638 Contracts or Memorandum of Agreements (MOAs) with water user groups for O&M are not excluded.

Section 203, which refers to "requirements and condition," includes the Commissioner of Reclamation. Section 2 also describes the Secretary of the Interior as acting through the Commissioner or Reclamation. We recommend amending these references since the Bureau of Reclamation (Reclamation)is not involved in the operation and maintenance of the 17 eligible irrigation projects, nor is Reclamation involved in the allocation of resources among funds created within the Treasury.In Section 203(2)(E), regarding the funding prioritization criteria/methodology, the BIA has been collecting Condition Assessment data that BIA has used over the past decade. The BIA is initiating irrigation project Modernization Studies, which will provide additional and very valuable decision-making information on how BIA would best rehabilitate the existing irrigation project infrastructure. We recommend adding language to Section 203 that allow the results of the Modernization Studies to be incorporated into the criteria/methodology used inthe implementation of S. 438.

Section 205 requires Tribal Consultation and Water User Input. As this Committee is aware, Tribal consultation can be a lengthy process, and we understand our responsibility to consult with affected tribes and water users for these 17 irrigation projects and to provide adequate notice to make such consultation meaningful, usually 30 days. In order to facilitate the consultation required under this Section, we recommend the timeframe be changed from "60 days" to a "not later than 120 days."

Section 206 of S. 438 provides a foundation for establishing which projects to prioritize and takes into consideration a reduced priority for those projects that have received funding by an "act of Congress" in the previous 15 year period.According to Section 206(b), the following projects have received funding under an "Act of Congress that expressly identifies the Indian irrigation project or the Indian reservation of the project to address the deferred maintenance, repair, or replacement needs of the Indian irrigation project:"

- Crow Indian Irrigation Project, Crow Tribe Water Rights Settlement Act of 2010, Public Law 111-291, signed into law on December 8th, 2010.
- Duck Valley Indian Irrigation Project, 2009 Omnibus Public Land Management Act (H.R. 146, 111th Congress)
- San Carlos Indian Irrigation Project – Indian Works, Arizona Water Settlements Act, 118 STAT. 3478 PUBLIC LAW 108–451—DEC. 10, 2004
- Navajo Indian Irrigation Project, Public Law No. 87-483, 76 Stat. 96 (1962) (NIIP Act) – receives annual appropriations for construction and maintenance activities.

Section 206(b) provides the "Priority"as first based on an Indian irrigation project(s) serving more than one Indian tribe within "an Indian reservation." Based on this requirement the followingirrigation project would receive funding priority under S. 438 over the first few years.

Priority List

The only Irrigation Project that serves more than one Indian tribe within an Indian reservation is:

- Wind River Indian Irrigation Project (Northern Arapahoe Tribe, Eastern Shoshone Tribe, Wind River Reservation)

The priorities identified next according to Sec. 203 "programmatic goals" to fulfill S. 438, and "critical maintenance needs" include the following Projects not listed in order of priorities:

- Blackfeet Indian Irrigation Project
- Colorado River Indian Irrigation Project
- Flathead Indian Irrigation Project
- Fort Belknap Indian Irrigation Project
- Fort Hall Indian Irrigation Project
- Fort Peck Indian Irrigation Project
- Pine River Indian Irrigation Project
- Pyramid Lake Indian Irrigation Project
- San Carlos Indian Irrigation Project – Joint Works
- Uintah Indian Irrigation Project
- Walker River Indian Irrigation Project
- Wapato Indian Irrigation Project

Irrigation Projects that serve only 1 Indian tribe within an Indian reservation but have received funds in last 15 years (if S. 438 enacted this year):

- Crow Indian Irrigation Project (received Settlement funds)
- Duck Valley Indian Irrigation Project (received Settlement funds)
- San Carlos Indian Irrigation Project – Indian Works (received Settlement funds)
- Navajo Indian Irrigation Project (receives annual appropriations)

Last September, we testified before this Committee that the BIA operates its irrigation projects consistent with numerous laws, regulations and policy guidance and many projectshave extensive and specific legislative histories. For example, specific statutory authorities require that BIA chargean assessment to both Indian and non-Indian customersfor O&M costs. Most of the 15revenue-generating projects receive little or no appropriated funds. Whenever possible and practical, BIA works to leverage cost-share opportunities with any other funding that is made available to tribes and water user organizations. Funding to maintain these systems must also compete for other pressing priorities in Indian Country.

Historically, BIA has not charged sufficient Operation, Maintenance & Rehabilitation (OM&R) rates to allow for adequate project maintenance and replacement. Over time, this has resulted in less maintenance accomplished and a steady increase in deferred maintenance. This contributed to critical reviews by the Office of Inspector General in the 1990's and the Government Accountability Office in 2006. The 2013 deferred maintenance estimate for BIA-owned irrigation facilities is approximately $600 million. Less clear is what should be the appropriate allocation of responsibility between the users and beneficiaries of these systems, particularly by non-tribal members, and the general taxpayer.

Over the past 9 years we have increased our O&M rates an average of 26% across all projects. We believe that rates are approaching levels to stem the growth of deferred maintenance, but the existing level of deferred maintenance is such that it may be difficult to address through increased O&M rates alone. To ensure the protection of the investments that would be provided by this Act, BIA will continue to evaluate the O&M rates assessed to irrigators while considering the local agricultural economies. The BIA irrigation projects are vital economic contributors to the local communities and regions where they are located. The BIA estimates that irrigated lands served by the15 BIA revenue generating irrigation projects add $490M in revenue and supports almost 10,000 jobs. This Administration supports investments in vital economic contributors and supports the goals of the bill, and we look forward to working with you to address the best means of doing so given current budget constraints and the ability of irrigation projects to financially sustain themselves in the long run.

This concludes my prepared statement. I will be happy to answer any questions you may have.

In: Indian Irrigation Projects
Editor: Christopher Flores
ISBN: 978-1-63483-261-8

Chapter 3

TESTIMONY OF CHARLES HEADDRESS, EXECUTIVE BOARD MEMBER, ASSINIBOINE AND SIOUX TRIBES OF FORT PECK RESERVATION. LEGISLATIVE HEARING ON "S. 438, THE IRRIGATION REHABILITATION AND RENOVATION FOR INDIAN TRIBAL GOVERNMENTS AND THEIR ECONOMIES ACT, THE IRRIGATE ACT"*

Good afternoon Chairman Barrasso and Vice-Chairman Tester. My name is Charles Headdress, Sr., and I am member of the Fort Peck Tribal Executive Board, the governing body of the Assiniboine and Sioux Tribes of the Fort Peck Reservation. I want to thank you for holding this hearing on S. 438, the Irrigation Rehabilitation and Renovation for Indian Tribal Governments and their Economies (IRRIGATE) Act. I also want to express my appreciation for two of the bill's co-sponsors: Senator Tester for the invitation to testify today; and Montana's junior Senator Steve Daines for his interest on this important subject and in serving on this important Committee.

The Fort Peck Reservation encompasses 2.1 million acres—over two thousand square miles—in remote northeastern Montana. The Assiniboine and

* This is an edited, reformatted and augmented version of testimony presented March 4, 2015 before the Senate Committee on Indian Affairs.

Sioux Tribes and individual Indian allottees own about 1 million acres of land on the Reservation. Nearly 10,000 people live on the Reservation, of which roughly two-thirds are Tribal members and non-member Indians.

The development of an irrigation project for the Fort Peck Reservation was an integral element of the plan and obligation that the federal government assumed when it established our Reservation. After our Reservation was created, the federal government, using military force, prohibited our people from leaving the Reservation to hunt the game on which we historically depended to meet all of our most basic needs. The government instead sought to have us engage in farming and ranching. But the Reservation resources were not sufficient to do this. After several years of drought and starvation among our people, the government recognized the need to develop irrigation so that we might, in fact, survive by agriculture.

The Fort Peck Irrigation Project was formally authorized by the Act of May 30, 1908. Importantly, pursuant to this 1908 Act, it was the Bureau of Reclamation that was charged with direct responsibility for materials, workmanship, and economy of construction of the irrigation system. Congress recognized that the Bureau of Indian Affairs did not have the capability of constructing the kind of irrigation project that was needed at Fort Peck. When Congress enacted this legislation it intended to provide the Fort Peck Tribes with the means to become self-supporting through the development of agricultural and grazing lands. Accordingly, the project was planned with the intent of irrigating up to 152,000 acres of land. Unfortunately, for a number of reasons associated with various failed federal policies – including allotment, removal of children from homes (which impeded the ability to run family farms, as there were no families), and the levying of construction debt against trust property – this never came to be.

Instead today, the Fort Peck Reservation Irrigation Project consists of two irrigation units: the Wiota Unit and Frazer-Wolf Point Unit. Together these units irrigate only 18,953 acres, approximately 12% of what was initially planned to serve my Reservation to meet the needs of my people. Out of these 18,953 acres, only 9,758 acres remain in trust, with the other 9,195 acres held in fee status. Some of the fee lands are owned by Tribal members.

The current condition of the Fort Peck Reservation Irrigation System is worse than poor. The national backlog of deferred maintenance on irrigation projects is in excess of $600 million. According to the BIA's 2014 Q4 Deferred Maintenance Report, the total backlog deferred maintenance for the Fort Peck Project is $12.7 million.

The impact of this deferred maintenance on the economy of the Fort Peck Tribes cannot be understated. Throughout the history of the Reservation the income generated by the farming and grazing has been a mainstay for the Tribes and Tribal members. Even in the years when there was an oil and gas boom for the Reservation, the revenue generated from grazing and agricultural leasing of the Tribes' trust lands was still approximately 30-50% of the Tribes' total trust income. In more recent years, where revenues from oil and gas have declined, the agricultural revenues are at the heart of funding Tribal government operations, programs, and services that are so critically needed by our people. The repair and restoration of the irrigation system is also key to creating jobs. The Fort Peck Reservation's unemployment rate has hovered above 50% for most of the last two decades. Poverty among our members remains at epidemic levels, as illustrated by the fact that more than 80% of our children are eligible for free or reduced-price school lunch. We have to do more to put our people to work and lift our children out of poverty. It is time for Congress to fulfill the original promise of the 1908 Act to make our Reservation self-supporting.

Thus, the Fort Peck Tribes stand in support of the IRRIGATE Act. This Act will address the $600 million tribal irrigation maintenance and repair backlog by allocating from the Reclamation Fund, $35 million each year from 2015 through 2036, into a new account in the Treasury called the Indian Irrigation Fund.

For those who might say this is not an appropriate use of the Reclamation Fund, we would like to correct them. The use of the Reclamation Fund to repair the Fort Peck Irrigation Project is entirely appropriate and is in fact, exactly the purpose for which the Reclamation Fund was established. The Fort Peck Project was originally developed as a Reclamation Project. The fact that the project could not generate revenues necessary to maintain itself is a consequence of Reclamation's failure to do its job right in the first place.

In this regard, the Tribes ask that the legislation be clarified to state that use of these funds to repair these Tribal irrigation projects is not a reimbursable expense. In the past, when money was appropriated to repair tribal irrigation projects, the Department deemed it to be reimbursable and levied additional assessments against the users. Currently, the unpaid construction debt on Fort Peck Reservation system is $7 million. Almost two decades ago, the Bureau of Reclamation did an analysis of the users' payment capacity and found it to be $15.50 per acre ($14.00 in O&M and $1.50 in construction repayment). Those figures would have meant it would have taken 250 years to satisfy the construction debt. Demanding repayment was not realistic then and it is not realistic now. We do not believe you intend the funds to be provided by the IRRIGATE Act to be added to the Project's construction debt, but given the Department's past positions, we urge that the legislation make clear that the funds provided are not to be subject to repayment by the users.

Moreover, to ensure the viability of the Fort Peck Project, we urge Congress to forgive the existing project debt. The amount of idle acreage continues to increase because landowners and potential lessees cannot afford to bring the debt current to get water delivered to the property. This both leaves the land idle, and decreases its value. As a consequence the Fort Peck Project is under-serving the intended project area. Even more troubling, we have heard from fractionated landowners within the Project that the government has become increasingly aggressive in seeking to recover Project debt. The Tribes at Fort Peck have been urging Congress to address this unfairness since 1993. While the Secretary has the authority to forgive this debt, our pleas have fallen on deaf ears. Congress has forgiven irrigation project debt in the past, including past debt for the Fort Peck Project, and we urge Congress to act now to forgive the current debt.

Beyond strengthening our tribal economy, repairing this project and addressing its debt would help preserve our resources. We are all aware of the need to use our resources efficiently, and there is no natural resource more precious to our people than water. This is especially true when the West is facing some of the worst droughts this Nation has ever experienced. By repairing and maintaining this Project, we will be ensuring that the water resources that the Fort Peck Tribes have fought so hard to protect are used wisely and efficiently.

In this regard, last Congress both the Chairman and the Vice-Chairman supported a similar bill in the Senate, the Rural Water Projects Completion Act, which was approved by the Senate Energy and Natural Resources Committee. Senator Daines introduced a companion bill during his tenure in the House. In addition to addressing the irrigation project maintenance and repair backlog, these bills would have created a mechanism to complete the several Rural Water Projects that have been authorized by Congress. These Projects include the Fort Peck Reservation Rural Water System and the Rocky Boy's North Central Project. The fate of our Reservations rests on the health of our people, and the health of our people depends on the water we drink. I know that Senator Tester knows this. There is probably no other United States Senator who cannot drink the water that comes from his kitchen sink, but I know that is the case for Senator Tester. Thus, I would urge the Committee as you take up this bill to follow the leadership of Senators Tester and Daines to amend it to include provisions that would also ensure that these rural water projects can be completed on time.

I would like to thank you for your time and interest in this vitally important matter, and I would be happy to answer any questions.

In: Indian Irrigation Projects
Editor: Christopher Flores
ISBN: 978-1-63483-261-8

Chapter 4

TESTIMONY OF MITCHEL T. COTTENOIR, TRIBAL WATER ENGINEER, SHOSHONE AND ARAPAHO TRIBES OF WIND RIVER RESERVATION. LEGISLATIVE HEARING ON "S. 438, THE IRRIGATION REHABILITATION AND RENOVATION FOR INDIAN TRIBAL GOVERNMENTS AND THEIR ECONOMIES ACT, THE IRRIGATE ACT"*

Chairman Barrasso, Vice Chairman Tester and members of the Committee, thank you for inviting me as a representative of the Eastern Shoshone and Northern Arapaho Tribes to appear before you today.

The condition of the Wind River Irrigation Project, as well as numerous other Bureau of Indian Affairs operated irrigation systems, is well documented. The Wind River Irrigation Project was authorized for construction in 1905, but was never completed. Since that time the project, under the operation of the BIA, has been neglected to the extent that the cost to rehabilitate and complete the system is estimated in the range of $30 - $90 million.

---

* This is an edited, reformatted and augmented version oftestimony presented March 4, 2015 before the Senate Committee on Indian Affairs.

Bureau of Indian Affairs Irrigation at the Wind River is significantly understaffed and the system is operated inefficiently with only minor necessary maintenance.

The BIA continues to not have a long term plan for rehabilitation of the Wind River Irrigation Project. Therefore, the Eastern Shoshone and Northern Arapaho Tribes, the Wind River Water Resource Control Board and the Office of the Tribal Water Engineer have undertaken a major rehabilitation effort to rehabilitate aging structures that are crucial to the operation of the system.

To further provide the required operational and maintenance needs of the system the Tribes have encouraged irrigators in the system to form water users associations. These associations have negotiated Cooperative Assistance Agreements (CAA) with the Bureau of Indian Affairs to assume the operation and maintenance of their designated portion of the system. A percentage of the irrigation assessment is returned to the association to provide funding for operating staff and needed maintenance. Under the CAA each association has seen a dramatic improvement the overall operation and maintenance of their part of the system compared to the past services provided by the Bureau of Indian Affairs.

It is hoped that each association can accumulate a rehabilitation fund to assist in the rehabilitation effort and that can be leveraged to acquire additional funding from sources such as the Wyoming Water Development Commission (WWDC). We have had some success in such efforts in recent years.

In addition, the Tribes have initiated an effort to assume the Operation and Maintenance responsibilities from the BIA under the Indian Self-determination Act PL 93-638. This action would empower the Tribes to operate the system more efficiently and effectively. Rehabilitation would become a priority rather than an afterthought. This effort has also been encouraged by Agency and Regional Level Bureau of Indian Affairs Water Resource management.

With these two strategies, the Bureau of Indian Affairs is eliminated from the equation. It leaves us with the responsibility to operate, maintain and rehabilitate the aging and deteriorating system on the Wind River.

The Eastern Shoshone and Northern Arapaho Tribes, the Wind River Water Resource Control Board and the Office of the Tribal Water Engineer strongly endorseS. 438, theIrrigation Rehabilitation and Renovation for Indian Tribal Governments and TheirEconomies Act" or the "IRRIGATE Act. The funds through this bill would provide for the much needed rehabilitation of the Wind River Irrigation Project that has for decades been neglected by the Bureau of Indian Affairs. The Eastern Shoshone and Northern Arapaho Tribes ask for your individual support in successfully moving the bill forward.

The Tribes have compiled a proven track record and have demonstrated the ability to move the rehabilitation effort forward for the benefit of not only Tribal members, but also our non-tribal neighbors. Funds that would become available to the Tribes through the IRRIGATE Act would enable the Tribes to continue this effort. The IRRIGATE Act funding could be utilized to leverage funding from the State of Wyoming and the WWDC. Tribal participation in this program was allowed for under the 2003 Wyoming State Legislature House Bill 144. In doing so, this could expedite the much needed rehabilitation and completion of the Wind River Irrigation Project which has so long been neglected by the Bureau of Indian Affairs.

In 2004 in an effort to facilitate the rehabilitation of the Wind River Irrigation Project, the Eastern Shoshone and Northern Arapaho Tribes through the efforts of the Wind River Water Resource Control Board applied to and were granted a $3.5M grant from the WWDC to aid in the rehabilitation of irrigation structures that were critical to the operation of the system. This State Appropriation was a 50% grant that required an additional $3.5M in matching funds before the State funds could be used. Through the efforts of the Wind River Water Resource Control Board in conjunction with the efforts led by Senator Mike Enzi, a Federal appropriation of $3.72M was secured in 2005 and 2006 as matching funds for the $3.5M in State funds.

These funding sources were utilized to rehabilitate 15 major structures that were crucial to the operation of the irrigation system. These structures include: the Johnstown and Lefthand Ditch diversion and waste-way structures on the Big Wind River, the Coolidge Canal – Trout Creek diversion structure, the Mill Creek – Ray Canal Crossing structure, the Ray Canal – South Fork of the Little Wind diversion structure, the Coolidge Canal – Little Wind diversion structure, Ray Canal 11C, 39C and 59C diversion structures, Coolidge Canal 14B diversion structure, the Sub-agency Canal – Little Wind River diversion structure, the North Fork of the Little Wind River diversion chute structure, and the Willow Creek and Meadow Creek diversion structures in the Crowheart area.

Incorporated in the design and construction of the Coolidge and Sub-agency structures are Fish Ladders. In addition to a Fish Ladder, a Fish Screen structure was also designed and constructed on Ray Canal. The fish passage will mitigate the loss of hundreds of thousands of fish to the irrigation system. The fish passage project was a combined effort among the Tribes, the US Fish and Wildlife Service, the Bureau of Indian Affairs, Trout Unlimited and the State of Wyoming.

The total cost of these 15 structures (Phase I of the Wind River Irrigation Rehabilitation Project) was $7,713,695.

Without the efforts of the Eastern Shoshone and Northern Arapaho Tribes through the Wind River Water Resource Control Board, Phase I of the rehabilitation process would not be occurring.

The Tribes and WRWRCB continue to pursue additional funds for the rehabilitation effort from the State of Wyoming through the WWDC. The Tribes have come to the table with $730,000 and have requested a matching grant in the amount of $1,482,121 from the Wyoming Water Development Commission. These funds will enable the Tribes to address the rehabilitation of structures identified on the Phase II priority list in Table 1.

Table 1. Wind River Irrigation Project Phase II Priority List

Ranking	Project Name	Total Cost in 2008 Dollars
1	Coolidge Canal Lateral 14B	$254,000
2	Dinwoody Diversion Structure	$552,000
3	Sub-agency Lateral 38 A	$180,000
4	Ray Canal Lateral 37 C System	$432,000
5	Coolidge Canal System	$150,000
6	Sub Agency Canal System	$282,000
7	North Fork Diversion Structure	$300,000
8	Mill Creek / 14B-44 System	$180,000
Total Cost		*$2,334,000*
	Estimated Cost in 2014 Dollars	$2,583,738
	As per the Bureau of Labor and Statistics Calculator	
Current Funding		
	Tribal Funding	$730,000
	Wyoming Water Development	$1,482,121
Total Funding		$2,212,121

The Tribes will continue this phased approach to the Rehabilitation Process. Additional phases and priority lists will be developed and added as funding is acquired.

Although the Tribes appreciate the financial support of the State of Wyoming, the funding only scratches the surface of what is necessary to bring the Wind River Irrigation Project up to the standards of non-Indian irrigation projects in close proximity to the Wind River Indian Reservation. The Tribes request the aid and assistance of both Senators Barrasso and Enzi and the

Senate Committee on Indian Affairs to help secure future funding for the ongoing rehabilitation of the Wind River Irrigation System. For this reason, the Eastern Shoshone and Northern Arapaho Tribes, the Wind River Water Resource Control Board and the Office of the Tribal Water Engineer again endorseS. 438, theIrrigation Rehabilitation and Renovation for Indian Tribal Governments and TheirEconomies Act" or the "IRRIGATE Act.

According to the GAO Report 06-314 dated February 2006, the Wind River Irrigation Project was authorized for construction in 1905 but construction was never completed.

The Wind River Irrigation Project is comprised of 3 storage facilities, 11 canals and 377 miles of canals and laterals. These facilities provide water to 38,300 acres of which 67% is Indian owned and 33% non-Indian owned.

According to the 1994 Natural Resource Consulting Engineers (NRCE) Project Assessment and Plan, no Project-wide rehabilitation of the delivery system has occurred since the 1930's. According to that study due to deferred maintenance over many years, 60% or 1200 structures were in need of repair or replacement and 45% or 190 miles of canals and laterals need repair or reconstruction. According to the study structure failures were routine resulting in the progressive loss of control of Project water and that catastrophic failure of segments of the delivery system was imminent. According to the 1994 NRCE Project Assessment and Plan due to the Project's current configuration, it only has 66 acres of irrigated land per mile of canal. In comparison, Midvale Irrigation District has over 160 acres per mile of canal. As a general guideline, the Bureau of Reclamation suggests that irrigation projects, in the region, need at least 140 acres of irrigated land per mile of canal to be economically self-sufficient. The study also stated that the resulting poor delivery performance had contributed to a progressive deterioration in crop quality and the water users' ability to pay assessments. It is apparent that the Wind River Irrigation System cannot be considered self-sufficient.

The condition of the Wind River Irrigation Project sadly continues to deteriorate. With the exception of the Tribes' Phase I Rehabilitation Project, little has changed since the 1994 NRCE Wind River Irrigation Project Assessment, the 2006 GAO-Report numbered 06-314 or the 2008 HKM Wind River Irrigation Project Engineering Evaluation and Condition Assessment. The $7,713,695of Phase I barely scratched the surface in addressing the needs as outlined in the 2008 HKM Wind River Irrigation Project Engineering Evaluation and Condition assessment where the estimated costs for needed replacement construction to be $69,640,000. According to the calculator on the Bureau of Labor and Statistics, something that cost $100 in 2008 cost $110.70 in 2014, which is a 10.7% increase; inflation raises those cost to approximately $77,091,500 in 2014.

Clearly something needs to be done. If funds are not made available to deal with the rehabilitation needed, the project will continue to lose water, and both the Indian and non-Indian people who rely on the project, as well as the fisheries impacted by the project, will all suffer.

In addition to the rehabilitation effort, in 2014 the Tribes successfully submitted Level II Phase II Storage Site Studyapplications to the WWDC. These studies will identify at least 2 suitable storage sites on each of the Big and Little Wind Rivers. The need for additional storage on the Wind River Reservation has been graphically demonstrated during drought years when irrigators have been shut off early in the summer months as early as the first or second week in July. These storage studies and the successful identification of storage sites will not only benefit Tribal irrigators but also all water users on the Wind/Big Horn River system.

What follows is a report on the Wind River Irrigation Rehabilitation Project. The photos graphically show what progress looks like, i.e., what we can jointly accomplish when we have the funding as well as demonstrate what happens when maintenance is deferred and the project is allowed to deteriorate.

In order for the rehabilitation effort to move forward, it will take a united effort from the Eastern Shoshone and Northern Arapaho Tribes, the Bureau of Indian Affairs, the Wyoming Water Development Commission, and our State and Federal Legislators.

Chairman Barrasso, Vice-Chairman Tester and members of the committee, the funding from this bill is simply vital to our efforts. We realize that only through our efforts, and yours, will this absolutely essential rehabilitation occur. Not only can we do this, we must do this.

Chairman Barrasso, the Eastern Shoshone and Northern Arapaho Tribes, the Wind River Water Resource Control Board and the Office of the Tribal Water Engineer strongly endorse S. 438, theIrrigation Rehabilitation and Renovation for Indian Tribal Governments and TheirEconomies Act" or the "IRRIGATE Act" as we did S. 715 when the Barrasso amendment was added to it in the previous Congress.

We also encourage members of this committee to do all that is their power to help in moving the IRRIGATE Act forward successfully. The Eastern

Shoshone and Northern Arapaho Tribes look forward to working closely with you now and in the future.

Your strong support of the Tribes and their efforts is of the utmost importance. Our efforts will bring much needed relief to both Tribal and non-Tribal irrigators on the Wind River Reservation.

If you have any questions or need additional information, please contact the Office of the Tribal Water Engineer at 307-332-6464.

Thank you for your time and consideration.

WIND RIVER IRRIGATION REHAB UPDATE BY THE OFFICE OF THE TRIBAL WATER ENGINEER

The Wind River Water Resource Control Board and the Office of the Tribal Water Engineer appreciates this opportunity to up-date the members of the United States Senate Committee on Indian Affairs on the progress of the Wind River Irrigation Rehabilitation Project.

In 2004 the Wyoming State Legislature appropriated $3,500,000 thru the Wyoming Water Development Commission to assist in the rehabilitation of the Wind River Irrigation Project. These funds were matched with two Federal appropriations in 2006 and 2007 totaling $3,722,500. The combined funding was utilized to rehabilitate irrigation structures in critical need of repair. To this date the total estimated cost of rehabilitating the Irrigation Project remains in the $90M range.

Ray Canal Diversion Structure and Fish Passage

During the 2010- 2011 and 2011-2012 construction seasons the following rehabilitation project structures were completed:

- Coolidge – Trout Creek Diversion Structure
- Johnstown Diversion Structure
- Lefthand Ditch Diversion and Waste Way Structures
- Ray Canal – Mill Creek Crossing Structure
- Ray Canal Diversion Structure
- Coolidge Canal Diversion Structure
- Ray Canal Fish Screen Structure
- Structures: 39-C, 11-C, 59-C, and 14-B

Total cost of rehabilitating these structures was $5,097,095.

In addition to replacing the Ray and Coolidge Canal diversion structures, fish ladders were installed on both diversion structures to enable aquatic life to migrate the stream above and below the diversion structures. A fish screen structure was also constructed in the Ray Canal down steam of the diversion structure. The fish screen will prevent the loss of thousands of fish to the Ray Canal irrigation system. The fish screen structure was built thru the collaborative efforts of various agencies. Funding partners for the fish passage and screens were: USF&W, Trout Unlimited, the Wyoming Wild Life Trust Fund, BIA Wildlife Resources Branch, the WWDC and the Eastern Shoshone and Northern Arapaho Tribes. These funding partners contributed an additional $720,760.

Ray Canal Fish Screen

The 2013- 2014 construction season brought this phase of the Wind River Irrigation Rehabilitation Project to a close. WWDC and Federal Funding have been depleted with the start of the 2014 irrigation season. Projects constructed during this period were:

- North Fork Chute
- Willow Creek Diversion Structure
- Meadow Creek Diversion Structure
- Sub-agency Diversion Structure

The total construction costs of these structures were $2,616,599.29.

Sub-agency Diversion Structure

Meadow Creek

Willow Creek

North Fork Chute

It should be noted this rehabilitation project has provided an economic boost to Fremont County, the State of Wyoming and the Wind River Indian Reservation. State and local contractors and sub-contractors have been utilized during the construction projects. These contractors include:

- Inberg-Miller Engineers (Riverton, WY)
- Lowham Walsh LLC (Lander)

- Dowl HKM (Lander)
- High Country Construction (Lander)
- 71 Construction (Riverton)
- Reiman Corp. (Cheyenne)

This phase of the Irrigation Rehab Project has barely scratched the surface of the overall needs of the Wind River Irrigation Project. As stated before the estimated cost of total rehabilitation is in the $90M range. The Office of the Tribal Water Engineer has developed a Phase II list of priorities for rehabilitation. This list and a cost estimate was submitted to the WWDC in November of 2014. We are currently awaiting Wyoming Legislative action on our funding application.

What follows are examples of conditions that currently exist on the Wind River Irrigation Project that need to be addressed in the Rehabilitation Effort.

In: Indian Irrigation Projects
Editor: Christopher Flores
ISBN: 978-1-63483-261-8

Chapter 5

INDIAN IRRIGATION PROJECTS: NUMEROUS ISSUES NEED TO BE ADDRESSED TO IMPROVE PROJECT MANAGEMENT AND FINANCIAL SUSTAINABILITY*

United States Government Accountability Office

WHY GAO DID THIS STUDY

The Department of the Interior's Bureau of Indian Affairs (BIA) manages 16 irrigation projects on Indian reservations in the western United States. These projects, which were generally constructed in the late 1800s and early 1900s, include water storage facilities and delivery structures for agricultural purposes. Serious concerns have arisen about their maintenance and management.

GAO was asked to examine (1) BIA's estimated deferred maintenance cost for its 16 irrigation projects, (2) what shortcomings, if any, exist in BIA's current management of its irrigation projects, and (3) any issues that need to be addressed to determine the long-term direction of BIA's irrigation program.

* This is an edited, reformatted and augmented version of a United States Government Accountability Office publication, No. GAO-06-314 dated February 2006.

What GAO Recommends

GAO recommends that BIA (1) provide the necessary level of technical support to project managers, (2) require project managers to meet at least twice annually with water users, and (3) conduct studies to determine the financial sustainability of the projects.

Although we requested comments from the Department of the Interior on our findings and recommendations, none were provided in time to be included as part of this report.

What GAO Found

BIA estimated the cost for deferred maintenance at its 16 irrigation projects at about $850 million for 2005, although the agency is in the midst of refining this estimate. BIA acknowledges that this estimate is a work in progress, in part, because some projects incorrectly counted new construction items as deferred maintenance. To further refine its estimate, BIA plans to hire engineering and irrigation experts to conduct thorough condition assessments of all 16 irrigation projects to correctly identify deferred maintenance needs and costs.

BIA's management of some of its irrigation projects has serious shortcomings that undermine effective decisionmaking about project operations and maintenance. First, under BIA's organizational structure, officials with the authority to oversee irrigation project managers generally lack the technical expertise needed to do so effectively, while the staff that have the expertise lack the necessary authority. Second, despite federal regulations that require BIA to consult with project stakeholders in setting project priorities, BIA has not consistently provided project stakeholders with the necessary information or opportunities to participate in project decisionmaking.

The long-term direction of BIA's irrigation program depends on the resolution of several larger issues. Of most importance, BIA does not know to what extent its irrigation projects are capable of financially sustaining themselves, which hinders its ability to address long-standing concerns regarding inadequate funding. Information on financial sustainability, along with accurate deferred maintenance information, are two critical pieces of information that are needed to have a debate on the long-term direction of

BIA's irrigation program. Once this information is available, the Congress and interested parties will be able to address how the deferred maintenance will be funded and whether entities other than BIA could more appropriately manage some or all of the projects.

Abandoned Car in Canal

Crumbling Concrete Structure

Source: GAO.

Deferred Maintenance at the Crow Irrigation Project (March 2005).

ABBREVIATION

BIA Bureau of Indian Affairs

* * *

February 24, 2006

The Honorable Conrad Burns
Chairman
Subcommittee on Interior and Related Agencies
Committee on Appropriations
United States Senate

The scarcity of water in the western part of the United States makes irrigation critical to the continued success of agricultural activities. There are over 100 irrigation works on Indian reservations primarily across the western United States. The Department of the Interior's Bureau of Indian Affairs (BIA), which is responsible for providing social and economic services to Indians, as well as managing land and natural resources held in trust by the United States for Indians, currently refers to these irrigation works as either "irrigation projects" or "irrigation systems." There are 16 irrigation projects where water users are charged for the operation and maintenance of the irrigation works by BIA. The remaining systems, on which BIA does not charge an annual operation and maintenance fee, are operated and maintained through a collaborative effort which generally involves other BIA programs, tribes, and water users.

The 16 irrigation projects, which were generally initiated in the late 1800s and early 1900s by the Department of the Interior, include water storage facilities and delivery structures for agricultural purposes. The projects were constructed as part of the federal government's Indian assimilation policy to foster agricultural opportunities and provide economic benefits to Indian communities. Over time, non-Indians began buying or leasing the land served by the projects for agricultural purposes, and project stakeholders evolved from Indian water users and the tribes within the reservations to include non-Indian water users as well. Many of the water users today are non-Indian.

A number of prior reports on BIA's irrigation projects have documented that the annual operations and maintenance fees have historically been set too low to cover the full cost of running the projects.[1] In addition, problems have been reported with collecting the fees that have been assessed. Because of insufficient funding, project maintenance has been consistently postponed, resulting in an extensive and costly list of deferred maintenance items. The kinds of such deferred maintenance range from repairing or replacing dilapidated irrigation structures to clearing weeds from irrigation ditches. In addition to the deferred maintenance, water users have expressed concern that BIA has been unresponsive in addressing the projects' ongoing operations and maintenance needs.

The 16 irrigation projects are located in 4 of BIA's 12 regions—Rocky Mountain, Northwest, Southwest, and Western. Management of these projects is decentralized, with local or regional BIA offices responsible for project management. Fourteen projects are overseen by local BIA agency superintendents, and the 2 largest projects are overseen directly by regional directors. The agency superintendents that oversee projects report to their respective regional director. BIA's irrigation and engineering experts, who provide technical assistance to the projects, are located in each region as well as in the BIA central Irrigation, Power, and Safety of Dams offices (central irrigation offices) located in Washington, D.C., and other BIA locations in the western United States. The regional irrigation staff and central irrigation office staff do not have line authority over the projects.

Federal regulations specify that in making judgments about the work and actions necessary for the proper operation, maintenance and administration of the projects, the official in charge "... consults with water users and their representatives, and with tribal council representatives, and seeks advice on matters of program priorities and operational policies."[2] Furthermore, the regulations state that "close cooperation between the Indian tribal councils, the project waters users and the Officer-In-Charge [of the project] is necessary and will be to the advantage of the entire project."[3] BIA's irrigation manual and handbook also contain language directing project staff to involve project stakeholders in the management of the projects.

In response to ongoing concerns about maintenance and management of the irrigation projects, in December 2003 Senator Conrad Burns and Congressman Dennis Rehberg, both of Montana, sponsored a town hall meeting with local water users and BIA officials to discuss problems at BIA's irrigation projects. In this context, we were asked to examine (1) BIA's estimated deferred maintenance cost for its 16 irrigation projects; (2) what shortcomings, if any, exist in BIA's current management of its irrigation projects; and (3) any issues that need to be addressed to determine the long-term direction of BIA's irrigation program.

To address the objectives of this report, we collected documentation on BIA's 16 irrigation projects from officials in BIA's central irrigation office, and we visited and collected information from each of BIA's four regional offices that oversee the 16 irrigation projects. We also visited 9 of the 16 projects, where we collected project-specific information from BIA officials and project stakeholders. We also met with and collected documentation from the Department of the Interior's Bureau of Reclamation, the primary agency responsible for irrigation management, for comparative purposes. Specifically, to examine estimated deferred maintenance costs, we reviewed BIA's lists of deferred maintenance items and cost estimates, and the methodology BIA used to develop these lists and estimates. We concluded that these data were sufficiently reliable for the purposes of this report based on a review of relevant controls, reliability tests, and interviews with agency officials about the collection and management of the data. We did not develop our own estimate of deferred maintenance. To determine what, if any, management shortcomings exist, we reviewed relevant federal regulations and agency guidance, and analyzed BIA-wide and project-specific management protocols and systems for the 9 projects we visited. Finally, to determine any issues needing to be addressed to determine the long-term direction of the projects, we reviewed prior studies on BIA's irrigation program and we discussed the long-term direction of the program with BIA irrigation officials and project stakeholders. A more detailed description of our objectives, scope and methodology can be found in appendix I. We performed our work between March 2005 and February 2006 in accordance with generally accepted government auditing standards.

RESULTS IN BRIEF

BIA has estimated the cost for deferred maintenance at its 16 irrigation projects at about $850 million for 2005, although the agency is in the midst of refining this estimate. BIA defines deferred maintenance as upkeep, such as removing weeds from irrigation ditches or repairing irrigation structures, that is postponed until some future time. As part of its ongoing strategy to develop a cost figure for the projects' total deferred maintenance, BIA had initially estimated this cost at approximately $1.2 billion in fiscal year 2004. This estimate was based, in part, on preliminary condition assessments of structures and equipment at each of the 16 irrigation projects using a scale of good, fair, poor, critical, and abandoned. The assessment of the structures consisted of visual inspections generally conducted by nonengineers. BIA acknowledged that the 2004 estimate would need revision largely for three reasons: the individuals who conducted the assessments did not have irrigation or engineering expertise, not all projects used the same methodology to develop their deferred maintenance estimates, and some projects incorrectly counted new construction items as deferred maintenance. To improve its estimate in 2005, BIA implemented a facilities management system designed to help projects track and continuously update deferred maintenance information. BIA technical experts from the central irrigation office conducted training for BIA irrigation projects on how to use this system, as well as how to correctly define deferred maintenance. Projects used this system to revise their list of deferred maintenance items and associated cost estimates in fiscal year 2005, resulting in a lower total deferred maintenance estimate of about $850 million. However, some projects continued to classify items as deferred maintenance when they were actually new construction, and some provided BIA with incomplete information. To further refine the cost estimate, BIA plans to hire experts in engineering and irrigation to conduct thorough condition assessments of all 16 irrigation projects every 5 years to identify deferred maintenance needs and costs. The first such assessment was completed in July 2005, with all 16 assessments expected to be completed by 2010.

BIA's management of some of its irrigation projects has serious shortcomings that undermine effective decisionmaking about project operations and maintenance. First, under BIA's organizational structure, in many cases, officials with the authority to oversee project managers' decisionmaking lack the technical expertise needed to do so effectively, while the staff who do have the expertise lack the necessary authority. The BIA regional directors, agency superintendents and deputy superintendents that oversee the projects do not generally have engineering or irrigation expertise and they rely heavily on the project managers to run the projects. However, this process breaks down when the project managers themselves do not have the expertise required for the position—that is, in cases in which BIA has had difficulty filling project manager vacancies and has, as a result, hired less qualified people. For example, at the Crow project in 2002, a project manager with insufficient expertise decided to repair a minor leak in a key water delivery structure by dismantling it and replacing it with a different type of structure. The new structure was subsequently deemed inadequate by BIA's irrigation experts, and the required reconstruction delayed water delivery by about a month. Furthermore, the BIA staff with such expertise—regional irrigation engineers and central irrigation office staff—have no authority over the 16 projects. A second serious management shortcoming involves the extent to which some projects involve stakeholders in decisionmaking. Despite federal regulations that require BIA to consult with project stakeholders in setting project priorities, BIA has not consistently provided the information or opportunities necessary for stakeholders—both Indian and non-Indian water users—to participate in decisionmaking about project operations and maintenance. For example, the Wapato Irrigation Project shares little information on its spending with stakeholders, and the Pine River Irrigation Project does not meet with its non-tribal stakeholders, limiting stakeholders' ability to have an impact on project decisions and BIA's ability to benefit from this input.

The long-term direction of BIA's irrigation program depends on the resolution of several larger issues. Of most importance, BIA does not know to what extent its irrigation projects are capable of financially sustaining themselves, which hinders its ability to address long-standing concerns regarding inadequate funding. The projects were constructed without consideration for whether they could generate adequate income to be self-supporting, yet since the 1960s many have been considered generally self-supporting through fees paid by water users. The future of BIA's irrigation program also depends on the resolution of how the deferred maintenance will be funded. BIA currently has no plan for how it will obtain funding to fix the deferred maintenance items. Regardless of the precise cost estimate for total deferred maintenance, funding deferred maintenance costs in the hundreds of millions of dollars will be a significant challenge in times of tight budgets and competing priorities. In the interim, the Congress has appropriated approximately $7.5 million for some of BIA's irrigation projects for fiscal year 2006. Finally, it might be more appropriate for other entities, including other federal agencies, tribes, and water users, to manage some or all of the projects. Given that BIA must balance irrigation management with its many other missions in support of Indian communities, such as providing education and law enforcement, it may be beneficial to consider whether others for whom irrigation is more of a priority or an area of expertise could better manage some of the projects. Successful management of the projects by other groups, however, would depend on the characteristics of each project and its stakeholders. For example, turning over projects to tribes may be better suited to projects where most of the water users are Indian, whereas turning over projects to water users would be better suited to projects where water users share similar interests and have a desire to organize into an irrigation district or association.

To improve the ongoing day-to-day management of the projects in the short-term, we are recommending that the Secretary of the Interior direct the Assistant Secretary for Indian Affairs to provide the necessary level of technical support for project managers who have less than the desired level of engineering qualifications and to adequately involve project stakeholders in the management of the projects. To address the long-term financial sustainability of the projects, we are recommending that the Secretary direct the Assistant Secretary for Indian Affairs to conduct studies to determine the extent to which projects are capable of sustaining themselves. Information on financial sustainability, along with accurate deferred maintenance information, are two critical pieces of information that are needed to have a debate on the long-term direction of BIA's irrigation program. Once this information is available, the Congress and interested parties will be able to address how the deferred maintenance will be funded and whether entities other than BIA could more appropriately manage some or all of the projects. Since how to fund the deferred maintenance and who should manage the projects are future policy issues for the Congress to decide in collaboration with all interested parties, we are not making any specific recommendations to address them. Although we requested comments from the Department of the Interior on our findings and recommendations, none were provided in time to be included as part of this report.

BACKGROUND

BIA's irrigation program was initiated in the late 1800s, as part of the federal government's Indian assimilation policy, and it was originally designed to provide economic development opportunities for Indians through agriculture. The Act of July 4, 1884, provided the Secretary of the Interior $50,000 for the general development of irrigation on Indian lands.[4] Over the years, the Congress continued to pass additional legislation authorizing and funding irrigation facilities on Indian lands.

BIA's irrigation program includes over 100 "irrigation systems" and "irrigation projects" that irrigate approximately 1 million acres primarily across the West. BIA's irrigation systems are non revenue-generating facilities that are primarily used for subsistence gardening and they are operated and maintained through a collaborative effort which generally involves other BIA programs, tribes, and water users. In contrast, BIA's 16 irrigation projects charge their water users an annual operations and maintenance fee to fund the cost of operating and maintaining the project.[5] Most of BIA's irrigation projects are considered self-supporting through these operations and maintenance fees. The 16 irrigation projects are located on Indian reservations across the agency's Rocky Mountain, Northwest, Southwest, and Western regions (see fig. 1).

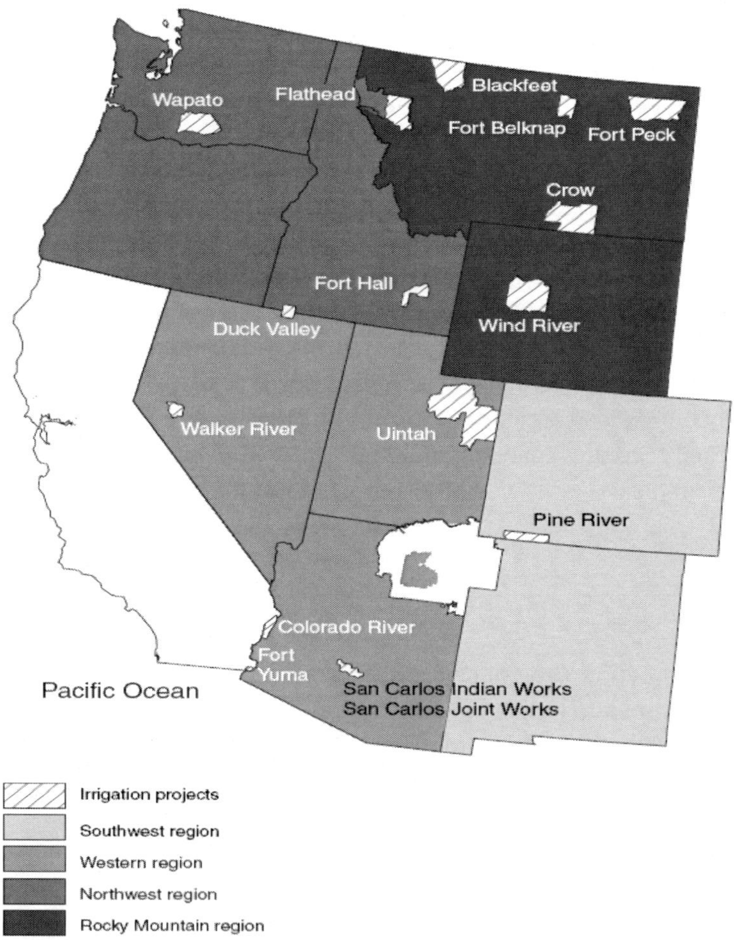

Sources: BIA; GAO.

Figure 1. Location of BIA's 16 Irrigation Projects by Region.

Table 1. Tribe(s) Served and Year Authorized for BIA's 16 Irrigation Projects

Irrigation project	Tribe(s) served	Year authorized
Blackfeet	Blackfeet Tribe of the Blackfeet Indian Reservation of Montana	1907
Colorado River	Colorado River Indian Tribes of the Colorado River Indian Reservation, Arizona and California	1867
Crow	Crow Tribe of Montana	1890
Duck Valley	Shoshone-Paiute Tribes of the Duck Valley Reservation, Nevada	[a]
Flathead	Confederated Salish & Kootenai Tribes of the Flathead Reservation, Montana	1904
Fort Belknap	Fort Belknap Indian Community of the Fort Belknap Reservation of Montana	1895
Fort Hall	Shoshone-Bannock Tribes of the Fort Hall Reservation of Idaho	1894
Fort Peck	Assiniboine and Sioux Tribes of the Fort Peck Indian Reservation, Montana	1908
Fort Yuma[b]	Quechan Tribe of the Fort Yuma Indian Reservation, California & Arizona	1904
Pine River	Southern Ute Indian Tribe of the Southern Ute Reservation, Colorado	[a]
San Carlos Indian Works	Gila River Indian Community of the Gila River Indian Reservation, Arizona	1924
San Carlos Joint Works	Gila River Indian Community of the Gila River Indian Reservation, Arizona	1924
Uintah	Ute Indian Tribe of the Uintah & Ouray Reservation, Utah	1906
Walker River	Walker River Paiute Tribe of the Walker River Reservation, Nevada	[a]
Wapato	Confederated Tribes and Bands of the Yakama Nation, Washington	1904
Wind River	Arapahoe Tribe of the Wind River Reservation, Wyoming and the Shoshone Tribe of the Wind River Reservation, Wyoming	1905

Source: GAO.

[a]No specific authorization date.

[b]The Fort Yuma Irrigation Project is operated and maintained by the Bureau of Reclamation. The operations and maintenance fees collected by BIA for the project are turned over to the Bureau of Reclamation.

BIA's management of the 16 irrigation projects is decentralized, with regional and local BIA offices responsible for day-to-day operations and maintenance. Table 1 provides the tribe or tribes served by each of the 16 irrigation projects along with the year each project was originally authorized.

The irrigation facilities constructed by BIA included a range of structures for storing and delivering water for agricultural purposes. Figure 2 highlights an example of the key structural features found on BIA's irrigation projects.

The beneficiaries of BIA's projects have evolved over time and at present are quite diverse. Over the years, non-Indians have bought or leased a significant portion of the land served by BIA's irrigation program. As a result, current water users on BIA's projects include the tribes, individual Indian landowners, non-Indian landowners, and non-Indian lessees of Indian lands. The extent of non-Indian landownership and leasing ranges significantly across BIA's irrigation projects (see table 2). For example, 100 percent of the land served by the Colorado River Irrigation Project is Indian owned, while only about 10 percent of the land served by the Flathead Irrigation Project is Indian owned.

Federal regulations and internal BIA guidance require that BIA collaborate with water users, both Indian and non-Indian, in managing the irrigation projects. For example, federal regulations state that close cooperation between BIA and water users is necessary and that the BIA official in charge of each project is responsible for consulting with all water users in setting program priorities.[6]

In addition, BIA's manual requires that BIA "provide opportunities for water user participation in matters relating to irrigation project operations" and that BIA's officer-in-charge "meet regularly with water users to discuss proposed [operation and maintenance] assessment rates ... [and] general operations and maintenance." Although BIA guidance does not define "regularly," BIA's Irrigation Handbook explicitly recommends that project staff meet at least twice annually to discuss work performed over the course of the year and allow for water user feedback and suggestions for the coming year. Furthermore, BIA's Irrigation Handbook states that, at a minimum, BIA should discuss annual project budgets and work plans with water users.

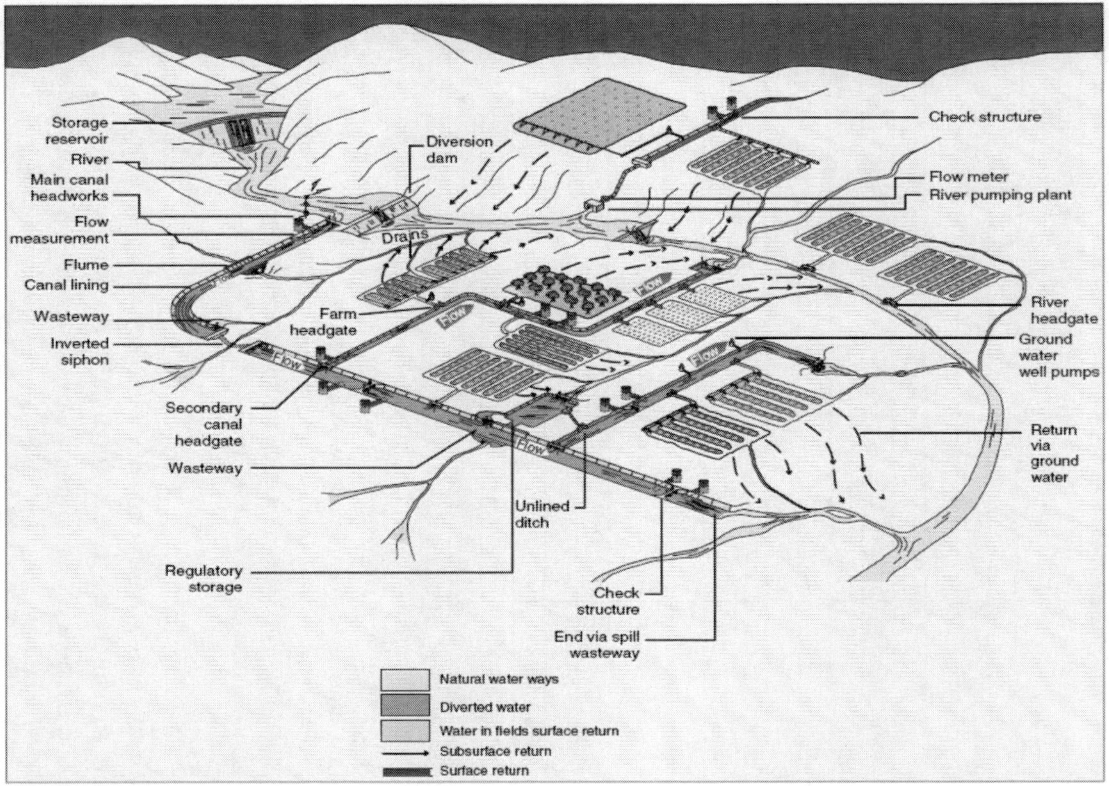

Source: Bureau of Reclamation, GAO.

Figure 2. Example of an Irrigation Project Operated by BIA.

Table 2. Land Ownership for BIA's 16 Irrigation Projects

Irrigation project	Total assessed acreage	Percentage of Indian owned land	Percentage of non-Indian owned land
Colorado River	79,350	100	0
Duck Valley	12,923	100	0
Fort Yuma[a]	7,524	100	0
San Carlos Indian Works	50,000[b]	99	1
Fort Belknap	9,900	92	8
Walker River	2,100	90	10
Pine River	11,855	85	15
Fort Hall	72,201	80	20
Wind River	38,300	67	33
Blackfeet	38,300	60	40
Wapato	96,443	60	40
Crow	38,900	56	44
Fort Peck	18,800	53	47
Uintah	62,200	52	48
San Carlos Joint Works	100,000[b]	50	50
Flathead	128,105	10	90
Total	716,901[b]	57	43

Source: GAO analysis of BIA data.

[a]The Fort Yuma Irrigation Project is operated and maintained by the Bureau of Reclamation. The operations and maintenance fees collected by BIA for the project are turned over to the Bureau of Reclamation.

[b]The acreage for the San Carlos Indian Works is also included in the acreage for the San Carlos Joint Works. In calculating the total acreage, the acreage for the San Carlos Indian Works is only counted once.

Since their inception, BIA's 16 irrigation projects have been plagued by maintenance concerns. Construction of the projects was never fully completed, resulting in structural deficiencies that have continually hindered project operations and efficiency. In addition, water users and BIA have reported that operations and maintenance fees provide insufficient funding for project operations. Due to insufficient funding, project maintenance has been consistently postponed, resulting in an extensive and costly list of deferred maintenance items. Such deferred maintenance ranges from repairing or replacing dilapidated irrigation structures to clearing weeds from irrigation ditches.

In addition, concerns regarding BIA's management of the projects have been raised for years, particularly in regard to its financial management practices. For example, problems concerning BIA's billing practices for its operations and maintenance fees have been raised by many, prompting independent review on more than one occasion. We and the Department of the Interior's Inspector General have both identified serious problems with the land use records BIA has used to develop its annual operations and maintenance bills.[7] In response, BIA instituted a new financial management system called the National Irrigation Information Management System, which has begun to address some of the billing errors. However, concerns still exist regarding the accuracy of the data in the billing system. The accuracy of some of the information in the irrigation billing system is dependant on the irrigation program receiving accurate and timely information from other BIA programs, such as land ownership and leasing information from BIA's Real Estate Services program.

In 2001, the Yakama tribe and individual tribal members filed appeals challenging the Wapato Irrigation Project's operation and maintenance fees for the pre-2000 and year 2000 bills. Furthermore, the Wapato Irrigation Project agreed to not send any bills to the tribe or its members since 2001. Although a settlement is under discussion, in the interim the Wapato Irrigation Project has not been able to collect about $2 million, annually, of its expected revenue.

BIA Estimates the Cost of Deferred Maintenance at about $850 Million, but the Estimate Is Being Refined

According to BIA's latest estimate, it will cost about $850 million to complete the deferred maintenance on all of its 16 irrigation projects; but this estimate is still being refined. BIA initially estimated its deferred maintenance costs at over $1 billion in fiscal year 2004, but acknowledged that this estimate was preliminary and would need to be revised largely because it incorrectly included new construction items and was developed by non-engineers. BIA revised this estimate downward in fiscal year 2005 based on the implementation of a new facilities management system. However, BIA plans to further refine this estimate since some projects continued to incorrectly count new construction items as deferred maintenance.

In 2004, BIA Initially Estimated Completing the Deferred Maintenance Would Cost Over $1 Billion

As part of its ongoing effort to identify the needs and costs of deferred maintenance on its 16 irrigation projects, BIA estimated in fiscal year 2004 that it would cost approximately $1.2 billion to complete all deferred maintenance. This initial estimate was based, in part, on preliminary condition assessments of irrigation structures and equipment for each of BIA's 16 irrigation projects. These preliminary condition assessments generally consisted of visual inspections to classify each project's structure and equipment using a scale of good, fair, poor, critical and abandoned based on the apparent level of disrepair. BIA staff then estimated how much it would cost to repair each item based on its condition classification.

Source: GAO.

Figure 3. Examples of Deferred Maintenance on BIA's Irrigation Projects (c. 2005).

BIA generally defines deferred maintenance as upkeep that is postponed until some future time.[8] Deferred maintenance varies from project to project and ranges from cleaning weeds and trees which divert water from irrigation ditches, to repairing leaky or crumbling check gates designed to regulate water flow, to resloping eroded canal banks to optimize water flow. Figure 3 shows examples of deferred maintenance on some of the irrigation projects we visited (clockwise from the upper left, figure 3 shows (1) a defunct check gate and overgrown irrigation ditch at the Fort Belknap Irrigation Project, (2) a cattle-crossing eroding a canal bank and impairing water flow at the Wind River Irrigation Project, (3) a crumbling irrigation structure at the Crow Irrigation Project, and (4) a check gate leaking water at the Colorado River Irrigation Project). For detailed information on key maintenance issues for each of the nine projects we visited, see appendix II.

BIA officials acknowledged that their fiscal year 2004 deferred maintenance estimate was only a starting point and that it needed to be revised for three key reasons: (1) the individuals who conducted the assessments were not knowledgeable about irrigation projects or infrastructure; (2) not all projects used the same methodology to develop their deferred maintenance cost estimates; and (3) some projects incorrectly counted new construction items as deferred maintenance.

- BIA's preliminary condition assessments were conducted by computer specialists, rather than by people with the expertise in irrigation or engineering needed to accurately assess project infrastructure. BIA contracted with geographic information system experts primarily to catalogue the structures on each project. These geographic information system experts also observed the condition of the structures they catalogued and classified the condition of each structure, based on the level of apparent disrepair, as part of the overall effort to inventory and map key structures on each project. Consequently, some items identified as being in "poor" condition may in fact be structurally sound but simply appear cosmetically dilapidated, whereas other structures classified as being in "good" condition may in fact be structurally dilapidated but appear cosmetically sound. For example, according to BIA staff at the Colorado River Irrigation Project, the recent repainting of certain check gates disguised severe rust and structural deterioration of key metal parts.
- BIA staff used inconsistent methodologies to develop the cost estimates for deferred maintenance. According to BIA staff, the deferred maintenance cost estimates were developed by different people, sometimes using different or unknown methodologies for assigning cost values to deferred maintenance items. For example,

some projects developed their own cost estimates and sent them to BIA's central office for inclusion in its overall figures, while BIA regional staff developed cost estimates for other projects based, in part, on information from BIA's preliminary condition assessments.

- Some projects incorrectly included new construction items as deferred maintenance. According to BIA, work that would expand a project or its facilities should not be categorized as deferred maintenance. Therefore, expanding an existing water delivery system or constructing a new building is not deferred maintenance. However, some projects incorrectly counted new construction items as deferred maintenance. For example, the Fort Hall Irrigation Project included increasing the capacity of its main canal for about $15.3 million, the Duck Valley Irrigation Project included building new canals for about $1.3 million, and the Flathead Irrigation Project included building a new warehouse for about $147,000.

In 2005, BIA Revised the Estimate Downward to about $850 Million, but It Is Still a Work in Progress

To improve the accuracy of its deferred maintenance estimate in 2005 and to help staff develop, track, and continuously update deferred maintenance lists and cost estimates, BIA implemented MAXIMO—a facilities management system linked to the geographic information system mapping inventory developed from its preliminary condition assessments.[9] Using data from MAXIMO, BIA revised its total deferred maintenance estimate for the irrigation projects downward to about $850 million for fiscal year 2005. Figure 4 shows the current deferred maintenance cost estimate for each of the 16 projects. In the summer of 2005, BIA technical experts from the central irrigation office conducted training for BIA irrigation projects on how to use MAXIMO to enter information on maintenance needs, and how to correctly define deferred maintenance. Projects used this system to revise their list of deferred maintenance items and associated cost estimates in fiscal year 2005. While MAXIMO is still being tailored to the needs of the irrigation program, its implementation generally standardized the process for identifying and calculating deferred maintenance among projects.

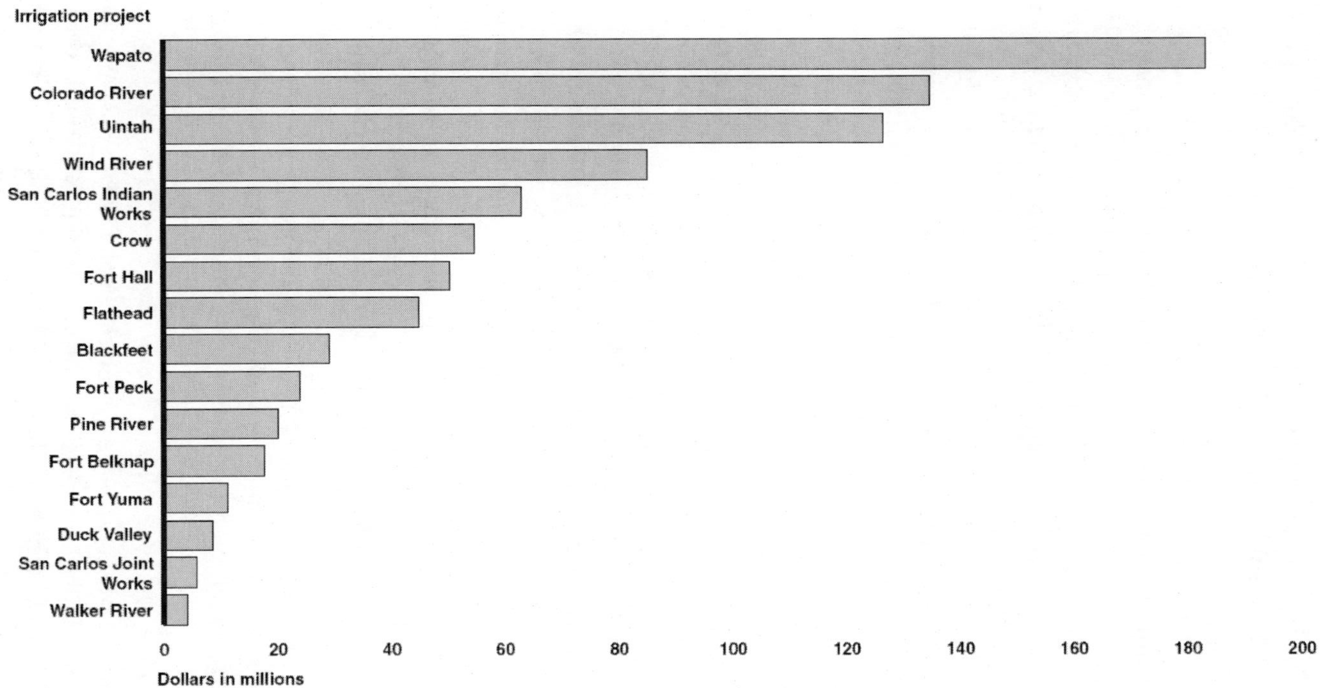

Source: GAO analysis of BIA data.

Figure 4. Fiscal Year 2005 Cost Estimate of Deferred Maintenance by Irrigation Project.

Despite the implementation of MAXIMO, BIA's fiscal year 2005 estimate of deferred maintenance is still inaccurate for the following reasons:

- *Some projects continued to incorrectly count certain items as deferred maintenance.* Despite training, some projects continued to incorrectly count certain items, such as new construction items and vehicles, as deferred maintenance. For example, the Fort Hall Irrigation Project included the installation of permanent diversion structures for about $2.1 million, the Wapato Irrigation Project included constructing reservoirs for about $640,000, and the San Carlos Indian Works Irrigation Project included building a new office for about $286,000. In addition, some projects included the cost of repairing vehicles or buying new ones in their deferred maintenance estimates, despite BIA's new guidance that such items are not deferred maintenance. According to BIA officials, while projects can consider the weed clearing postponed due to broken vehicles as deferred maintenance, the delayed repair of the vehicle itself is not deferred maintenance. For example, the Wind River Irrigation Project included an excavator vehicle for about $500,000 and the Crow Irrigation Project included dump trucks for about $430,000.
- *Some projects provided BIA with incomplete information.* According to BIA officials, some projects did not do thorough assessments of their deferred maintenance needs, and some may not be including legitimate deferred maintenance items, such as re-sloping canal banks that have eroded by crossing cattle or overgrown vegetation. Moreover, both the Walker River and the Uintah Irrigation Projects failed to provide information detailing their deferred maintenance costs, and several projects lumped items together as "other" with little or no explanatory information other than "miscellaneous"—accounting for almost one-third of BIA's total deferred maintenance cost estimate for its irrigation projects (see fig. 5).
- *BIA made errors when compiling the total deferred maintenance cost estimates.* For example, BIA inadvertently double-counted the estimate provided by the Colorado River Irrigation Project when compiling the overall cost estimate, according to BIA officials. Additionally, BIA officials erroneously estimated costs for all structures, such as flumes and check gates, based on the

full replacement values even when items were in good or fair condition and needed only repairs. These structures account for over one-third of BIA's total deferred maintenance estimate (see fig. 5).

While the inclusion of incorrect items and calculation errors likely overestimate BIA's total deferred maintenance costs, the incomplete information provided by some projects may underestimate total costs.

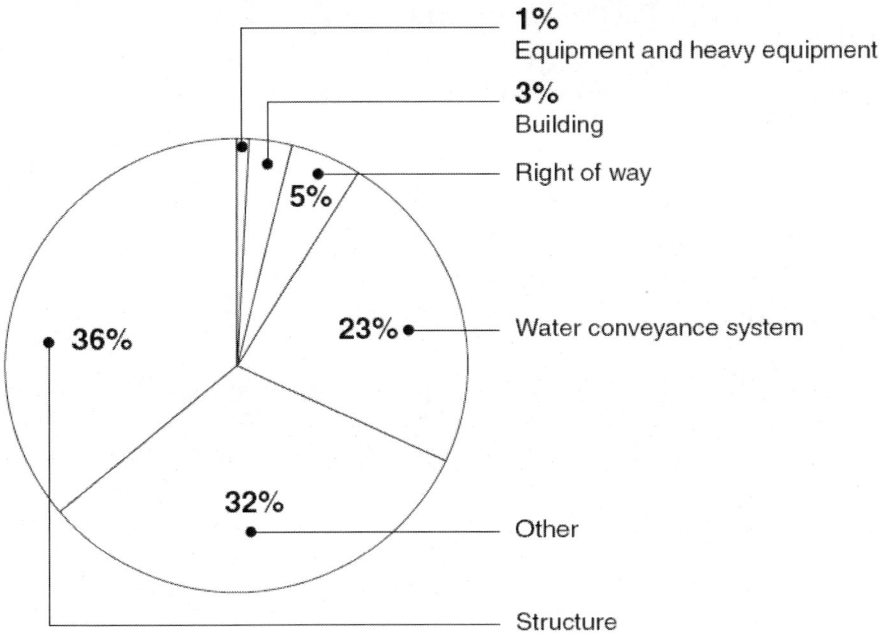

Source: GAO analysis of BIA data.

Figure 5. Fiscal Year 2005 Cost Estimate of Deferred Maintenance by Type.

To further refine its cost estimate and to develop more comprehensive deferred maintenance lists, BIA plans to hire experts in engineering and irrigation to periodically conduct thorough condition assessments of all 16 irrigation projects to identify deferred maintenance needs and costs. According to BIA officials, these thorough condition assessments are expected to more accurately reflect each project's actual deferred maintenance, in part because experts in engineering and irrigation who can differentiate between structural and cosmetic problems will conduct them. These assessments will also help BIA prioritize the allocation of potential funds to complete deferred maintenance items because they will assign a prioritization rating to each deferred maintenance item based on the estimated repair or replacement cost as well as the overall importance to the project. The first such assessment was completed for the Flathead Irrigation Project in July 2005,[10] and BIA plans to reassess the condition of each project at least once every 5 years, with the first round of such condition assessments completed by the end of 2010.

Shortcomings in BIA's Management of Some Irrigation Projects Undermine Effective Decisionmaking

BIA's management of some of its irrigation projects has serious shortcomings that undermine effective decisionmaking about project operations and maintenance. Under BIA's organizational structure, in many cases, officials with the authority to oversee project managers' decisionmaking lack the technical expertise needed to do so effectively, while the staff who do have the expertise lack the necessary authority. In addition, despite federal regulations that require BIA to consult with project stakeholders in setting project priorities, BIA has not consistently provided the information or opportunities necessary for stakeholders— both Indian and non-Indian water users—to participate in decisionmaking about project operations and maintenance. (See appendix II for detailed information on key management concerns at each of the nine projects we visited.)

In Many Cases, BIA Officials with Oversight Authority Lack Expertise, While Those with Expertise Lack Authority

Under BIA's organizational structure, in many cases, officials with the authority to oversee project managers' decisionmaking lack the expertise needed to do so effectively, while the staff who do have the expertise lack the necessary authority to oversee project managers' decisionmaking. BIA regional directors, agency superintendents, and agency deputy superintendents who oversee the projects do not generally have engineering or irrigation expertise, and they rely heavily on the project managers to run the projects. (See fig. 6 for an organizational chart showing the lines of authority for providing oversight of a typical BIA irrigation project.) Of the nine projects we visited, only two had managers at the regional or agency levels who are experts in irrigation or engineering. At the same time, BIA staff with the irrigation and engineering expertise— regional irrigation engineers and central irrigation office staff—have no authority over the 16 projects under BIA's current organizational structure. Consequently, key technical decisions about project operations and maintenance, such as when or how to repair critical water delivery infrastructure, do not necessarily get the technical oversight or scrutiny needed.

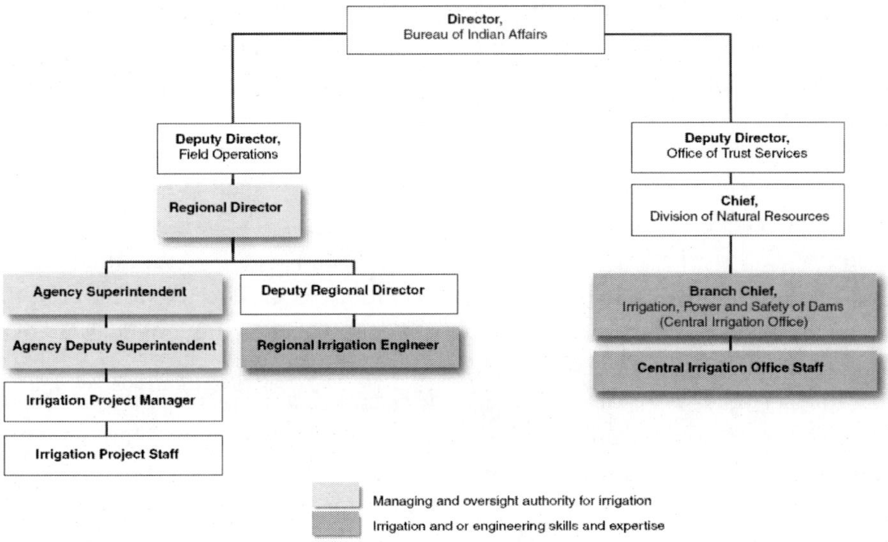

Source: GAO.

Figure 6. Misalignment of Expertise and Authority for a Typical BIA Irrigation Project.

This organizational structure and reliance on the project managers breaks down when the person managing the project lacks the expertise required for the position—that is, in cases in which BIA has had difficulty filling project manager vacancies and has, as a result, hired less qualified people or has the agency deputy superintendent temporarily serving in the project manager position. Of the nine projects we visited, four lacked project managers for all or part of the 2005 irrigation season and five project managers were experts in engineering or irrigation.

The GAO *Internal Control Management and Evaluation Tool* recommends that federal agencies analyze the knowledge and skills needed to perform jobs appropriately and provides guidance on organizational structure and identification of potential risks to the agency in that structure.[11] Specifically, it recommends that adequate mechanisms exist to address risks—such as the risks associated with staff vacancies or hiring less qualified staff.

When the project manager is under-qualified and unchecked by managers who heavily rely on his or her decisionmaking, the potential for adverse impacts on the operations and maintenance of an irrigation project increases. For example, at the Crow Irrigation Project in 2002, a project manager with insufficient expertise decided to repair a minor leak in a key water delivery structure by dismantling it and replacing it with a different type of structure. The new structure was subsequently deemed inadequate by BIA's irrigation experts, and the required reconstruction delayed water delivery by about a month. In addition, at the Blackfeet Irrigation Project in 2000, the accidental flooding and subsequent erosion of a farmer's land was inadequately addressed by project and agency management who decided to use a short-term solution over the objections of the regional irrigation engineer, who lacked the authority to override the project manager and agency superintendent's technical decision, despite their lack of expertise. At the time of this report, the regional irrigation engineer continues to negotiate the implementation of a long-term and technically sound solution.

Furthermore, BIA lacks protocols to ensure that project managers consult with, or get input from, BIA's technical experts before implementing technically complex decisions about project operations and maintenance, further exacerbating problems and undermining management accountability. For example, in the 2002 incident at the Crow Irrigation Project discussed above, the project manager was not required to consult with, notify, or get approval from either the regional irrigation engineer or central irrigation office staff, despite his lack of expertise and the complexity of the flume replacement project he undertook.

According to BIA officials, if the project manager had consulted an engineer, his plan to replace the flume with two small culverts would have been rejected before work began because it was technically insufficient and would not have been completed before the start of the approaching irrigation season.

BIA Has Not Consistently Provided Information and Opportunities for Stakeholders to Participate in Setting Project Priorities

A second serious management shortcoming is the extent to which some projects involve water users in decisionmaking. Federal regulations, as well as BIA guidance, call for involving project stakeholders—that is, tribal representatives as well as both Indian and non-Indian water users—in the operations and maintenance of each project. Specifically, federal regulations state that BIA is responsible for consulting with all water users in setting program priorities; BIA's manual requires that BIA provide regular opportunities for project water users to participate in project operations; and BIA's Irrigation Handbook recommends that BIA meet at least twice a year with project water users to discuss project budgets and desired work.

Despite such requirements and recommendations, BIA has not consistently provided the opportunities or information necessary for water users to participate in such decisionmaking about project operations and maintenance. The frequency of meetings between BIA and its project water users varied considerably on the nine projects we visited, from rarely (generally zero meetings per year), to periodically (generally more than one meeting per year), to regularly (generally more than three meetings per year), as shown in figure 9. For example, both the Blackfeet and Colorado River Irrigation Projects hold regular meetings with both tribal and individual water users, with meetings held quarterly at the Blackfeet Irrigation Project and monthly at the Colorado River Irrigation Project. In contrast, BIA officials on the Pine River Irrigation Project do not meet with any non-tribal water users, and BIA officials at the Fort Belknap Irrigation Project have held few water users meetings in recent years. There was no meeting with water users at the Fort Belknap Irrigation Project to kick-off the 2005 irrigation season because the project manager position was vacant, worsening an already adversarial relationship between water users and BIA, according to water users and a local government official. Also, BIA officials on the Crow Irrigation Project have no regularly scheduled meetings with either the tribe or individual water users and, in fact, failed to send a single representative to the meeting it called in 2005 for water users to voice their concerns about project management and operations.

	BIA communication with					
	Tribes			Individual water users[a]		
Irrigation project	Regular meetings	Periodic meetings	Rare meetings	Regular meetings	Periodic meetings	Rare meetings
Blackfeet	■			■		
Colorado River	■			■		
Crow			■			■
Fort Belknap			■			■
Pine River	■					■
San Carlos Indian Works		■			■	
San Carlos Joint Works		■			■	
Wapato	■				■	
Wind River		■		■		

Source: GAO.

[a]Individual water users include Indians and/or non-Indians, depending on the project.

Figure 7. Opportunities for Water Users to Meet with BIA Varies by Project.

In addition to a lack of regular meetings with all project water users, BIA has not consistently shared the type of information about project operations and finances that water users need to meaningfully participate in project decisionmaking. Although BIA officials at the Colorado River Irrigation Project share information on their budgets with water users and work collaboratively with water users to develop annual work priorities in accordance with BIA's *Irrigation Handbook*, not all projects we visited provide or solicit this type of information. For example, BIA staff at the Wapato Irrigation Project does not solicit water users' input on project priorities or share information on the project's budget, according to water users we spoke with, and BIA officials at the Crow Irrigation Project do not share this type of critical information. However, some of the projects we visited have recently begun to share information on project spending and involve project water users in developing project priorities, despite not doing so historically. For example, the project management at the Blackfeet Irrigation Project began sharing budget information with its water users during the 2005 season, and the new project management at the Fort Belknap Irrigation Project stated that they plan on involving project water users in setting project priorities in the 2006 season.

Moreover, although some project managers and their staff are approachable and responsive on an individual basis, according to water users on some projects we visited, others stated that project management on some of BIA's irrigation projects were generally inaccessible and nonresponsive. For example, BIA officials acknowledged that a former project manager at the Blackfeet Irrigation Project told water users to sue BIA to get information on project decisionmaking. In addition, some expressed concerns that BIA is less responsive to non-Indians because BIA's mission does not specifically include non-Indians. Consequently, some non-Indian water users have opted to go directly to their congressional representatives to raise their concerns. For example, non-Indian water users at the Wapato Irrigation Project have sought congressional intervention on several occasions to help compel BIA staff to disclose information about project finances, such as information related to proposed operations and maintenance fee debts and data on project land not being billed for operations and maintenance. In addition, Senator Conrad Burns and Congressman Dennis Rehberg of Montana co-sponsored a town hall meeting in 2003 to provide local water users an opportunity to voice project concerns to BIA officials. Requests by non-Indian water users for project management and regional staff to address the lack of water delivery at the Crow Irrigation Project during the month of August 2005 went largely unanswered by BIA, resulting in congressional intervention. Such lack of access and communication about project operations limits the ability of water users to have an impact on project decisions as well as the ability of BIA to benefit from this input.

Long-Term Direction of BIA's Irrigation Program Depends on Resolution of a Number of Larger Issues

The long-term direction of BIA's irrigation program depends on the resolution of several larger issues. Of most importance, BIA does not know the extent to which its irrigation projects are capable of financially sustaining themselves, which hinders its ability to address long-standing concerns regarding inadequate funding. The future of BIA's irrigation program also depends on the resolution of how the deferred maintenance will be funded. BIA currently has no plans for how it will obtain funding to fix the deferred maintenance items, and obtaining this funding presents a significant challenge in times of tight budgets and competing priorities. Finally, it might be more appropriate for other entities, including other federal agencies, tribes, and water users, to manage some or all of the projects.

The Extent to Which Projects Are Capable of Sustaining Themselves Is Unknown

BIA does not know the extent to which Indian irrigation projects are capable of sustaining themselves. Reclamation law and associated policy require the Department of the Interior's Bureau of Reclamation to test the financial feasibility of proposed projects comparing estimated reimbursable project costs with anticipated revenues. The Bureau of Reclamation then uses these reimbursable cost estimates to negotiate repayment contracts with water users, where appropriate. In contrast, Indian irrigation projects were authorized to support Indian populations residing on reservations without regard to whether the projects could be financially self-sustaining. As a result, neither the Congress nor project stakeholders have any assurance that these projects can sustain themselves. For example, a comprehensive 1930 study of BIA's irrigation program concluded that the Blackfeet and Fort Peck Irrigation Projects should be abandoned. Specifically, the report noted, "[a]fter a very careful study of all the available data relating to these projects, including a field examination, we are firmly convinced that any further attempts to rehabilitate and to operate and maintain these projects ... can result only in increasing the loss that must be accepted and sustained by the Government. *Adequate preliminary investigations and studies to which every proposed project should be subjected, in our opinion, would have condemned ... these ... projects as unfeasible*."[12] [Emphasis added]

Despite this lack of information on the overall financial situation for each of the projects, in the early 1960s BIA classified more than half of its 16 projects as fully self-supporting, on the basis of annual operations and maintenance fees they collected from water users. These self-supporting projects do not receive any ongoing appropriated funds. These projects are subject to full cost recovery despite the absence of financial information to demonstrate that the water users could sustain this financial burden. The Blackfeet and Fort Peck Irrigation Projects were two of the projects classified as fully self-supporting. While the specific financial situations for the Blackfeet and Fort Peck Irrigation Projects have likely changed since the 1920s, BIA does not know if these projects, or any of the other Indian irrigation projects, are financially self-supporting.

The heavy reliance on water users to sustain these projects has created ongoing tension between the water users and BIA. Some water users have complained to BIA that they cannot afford the operations and maintenance fees and they pressure BIA to keep the fees as low as possible. The Bureau of Reclamation recently conducted a study of the Pine River Irrigation Project and concluded that some of the water users could not conduct a profitable farming operation with the 2005 operations and maintenance fee of $8.50 per acre. BIA has not responded to the Bureau of Reclamation study, and in October 2005 BIA proposed doubling the rate to $17.00 per acre for the 2006 irrigation season even though water users claim that they cannot afford to pay a higher fee.[13] The operations and maintenance fee has been set at $8.50 at the Pine River Irrigation Project since 1992 and, according to BIA officials, the collections do not provide adequate funds to properly operate and maintain the project. As a result, BIA estimates that the deferred maintenance at the project has grown to over $20 million. Without definitive information on the financial situation of each project, BIA cannot determine what portion of project operations and maintenance costs can be reasonably borne by the water users and to what extent alternative sources of financing, such as congressional appropriations, should be pursued.

There Is No Plan for How to Fund the Deferred Maintenance

Despite the estimated $850 million in deferred maintenance and the degree to which it impedes ongoing operations and maintenance at BIA's irrigation projects, BIA currently has no plan for funding the list of deferred maintenance items. Funding deferred maintenance costs in the hundreds of millions of dollars will be a significant challenge in times of tight budgets and competing priorities. Nonetheless, officials stated that the agency has made little effort to identify options for funding the deferred maintenance. BIA acknowledges that income from ongoing operations and maintenance fees would likely be inadequate to cover the deferred maintenance, yet the agency has done little to identify alternative means of funding. According to officials, BIA has not asked the Congress for supplemental funding to cover the deferred maintenance. For example, water users report that the $7.5 million appropriated for BIA's irrigation projects for fiscal year 2006 resulted from lobbying by concerned water users, not from BIA's efforts.[14] To date, BIA has primarily focused on developing and refining an accurate estimate of the cost to fix the deferred maintenance items. While developing an estimate of the projected cost is important, BIA officials believe that the agency also needs to develop a plan for ultimately funding the deferred maintenance.

Developing a plan for funding the deferred maintenance is complicated by competing priorities and a crisis-oriented management style that complicates preventative maintenance, according to BIA officials. The current state of disrepair of most of the irrigation projects results in frequent emergency situations concerning project operations and maintenance. As a result, BIA irrigation staff spends a significant amount of its time addressing emergency maintenance situations, to the detriment of other maintenance needs that are essential to sustaining the projects over the long term. As a result of this "crisis-style" management, BIA has limited time to devote to non-emergency issues such as the list of deferred maintenance items. Furthermore, this "crisis-style" management prevents BIA from devoting adequate time to preventative maintenance. For example, irrigation staff at Wind River Irrigation Project stated that making "band-aid" emergency repairs on a regular basis prevents them from addressing long-standing deferred maintenance needs, as well as from conducting strategic improvements that would help sustain the project over the long term.

It Might Be More Appropriate for Other Entities to Manage Some or All of the Projects

It may be beneficial to consider whether other groups for whom irrigation is a priority or an area of expertise could better manage some of the irrigation projects, including other federal agencies, Indian tribes, and water users. BIA must balance its irrigation management responsibilities with its many other missions in support of Indian communities. As the federal agency charged with supporting Indian communities in the United States, BIA's responsibility is to administer and manage land and natural resources held in trust for Indians by the U.S. government. Administration and management of these trust lands and resources involves a wide variety of responsibilities, including law enforcement, social services, economic development, education and natural resource management. Given the multitude of responsibilities that BIA must balance, there are inherent limits on the resources and knowledge that BIA is able to devote to any one program. As a result of these limitations and competing demands, officials report that irrigation management is not a priority for BIA. The fact that many water users on the irrigation projects are now non-Indian may further encourage BIA to prioritize and devote more resources to other programs before irrigation management.

Successful management of the irrigation projects by other groups would depend on the unique characteristics of each project and its water users. Potential groups who may be able to assume management for some irrigation projects or portions of some irrigation projects include the following:

- *The Bureau of Reclamation.* As the federal agency charged with managing water in the western United States, the Bureau of Reclamation has extensive technical experience in managing irrigation projects and has served in a technical or advisory capacity to BIA's irrigation staff. Furthermore, efforts have been made in the past to turn over some BIA irrigation projects to the Bureau of Reclamation and the Fort Yuma Irrigation Project is currently operated by the Bureau of Reclamation. In addition, the Bureau of Reclamation utilizes management practices for its irrigation projects that maximize information sharing and collaboration with water users. For example, in contrast to BIA, the Bureau of Reclamation delegates responsibility for much of the day-to-day operations and maintenance on its irrigation projects to irrigation districts, which are organized groups of water users.
- *Indian Tribes.* Officials report that some of the tribes have staff with extensive knowledge of irrigation and water management, as well as technical training. Some tribes stated that they have a vested interest in seeing their respective projects succeed, and they would like to assume direct responsibility for their reservation's irrigation project, assuming the deferred maintenance items are fixed before the turnover occurs. Turning over some of the BIA projects to Indian tribes would be an option where tribes have the management and technical capability to assume responsibility for an irrigation project.

- *Water Users*. Water users have extensive familiarity with the day-to-day management of the projects and in some cases already handle many day-to-day operations and maintenance activities. For example, the Crowheart Water Users Association, a group of water users at the Wind River Irrigation Project, have successfully assumed responsibility for most of the maintenance needs on their portion of the project. In exchange for their efforts, BIA refunds to the Crowheart Water Users Association 50 percent of their annual operation and maintenance fees. Through this arrangement, the Crowheart Water Users Association believes it has been able to more effectively address maintenance needs and increase project efficiency. Turning over some of the BIA projects to water users would be an option where water users share similar interests and have positive working relationships, as well as the desire to organize an irrigation district or association.

Any successful alternative management option would have to consider the sometimes disparate interests and priorities among water users. In some cases, a combination of the various alternative management options may be beneficial and feasible. This type of arrangement is currently being considered for the Flathead Irrigation Project, where BIA is currently in the process of turning over the operation and management of the project to a collaborative management group that may include the tribe, individual Indian water users, and non-Indian water users. However, regardless of the alternative management option, water users and tribal officials repeatedly stated that they would not be willing or able to take over project operations and maintenance unless the deferred maintenance had already been addressed or adequate funding was available to address the deferred maintenance needs.

CONCLUSION

Since BIA historically has not had adequate funds to operate and maintain the projects, the projects are in a serious state of disrepair. BIA is in the process of implementing its plan to develop an accurate list and estimate of the deferred maintenance needs for each project. However, some of the projects also have day-to-day management shortcomings regarding technical support and stakeholder involvement that need to be addressed. BIA's decentralized organizational structure combined with the difficulty in attracting and retaining highly qualified project managers at remote Indian reservations led to some poor decisionmaking at some of the projects. It is critically important that project managers, especially those with less than desirable qualifications, have the necessary level of technical support to prevent poor decisions from being made in the future.

A lack of adequate stakeholder involvement at some projects has also seriously undermined project accountability. Unlike most other BIA programs, the operations and maintenance of the irrigation projects are funded almost entirely by the project beneficiaries—the water users, many of whom are non-Indian. Consequently, BIA is accountable to these water users and these water users expect to have an active voice in project operations and maintenance. Some projects have not fulfilled their obligations to regularly meet with project stakeholders, creating an adversarial environment in which BIA and project water users do not trust each other. This failure to involve stakeholders in the management of their own projects means that BIA does not benefit from water user expertise and has resulted in widespread feelings that BIA is non-responsive and evasive, alienating many water users who feel disenfranchised. Moreover, this failure has limited the ability of stakeholders to hold BIA accountable for its decisions and actions.

In addition to some shortcomings with BIA's ongoing day-to-day management of some of the projects, we also found that information on the financial sustainability of the projects is needed to help address the longterm direction of BIA's irrigation program. BIA's 16 irrigation projects were generally built in the late 1800s and early 1900s to further the federal government's Indian policy of assimilation. The government made the decision to build these projects to support and encourage Indians to become farmers. This decision was generally not based on a thorough analysis designed to ensure that only cost effective projects were built. As a result, the financial sustainability of some of the projects has always been questionable, ultimately creating tension between BIA and its water users. BIA is under constant pressure to raise annual operations and maintenance fees to collect adequate funds to maintain the projects, while many water users contend that they do not have the ability to pay higher fees. Without a clear understanding of the financially sustainability of the projects, BIA does not know whether it is practical to raise operation and maintenance fees, or whether alternative sources of financing should be pursued. Information on financial sustainability, along with accurate deferred maintenance information, are both critical pieces of information needed to have a debate on the long-term direction of BIA's irrigation program. Once this information is available, the Congress and interested parties will be able to address how the deferred maintenance will be funded and whether entities other than BIA could more appropriately manage some or all of the projects.

RECOMMENDATIONS FOR EXECUTIVE ACTION

We recommend that the Secretary of the Interior take the following three actions.

To improve the ongoing management of the projects in the short-term, we recommend that the Secretary direct the Assistant Secretary for Indian Affairs to

- provide the necessary level of technical support for project managers who have less than the desired level of engineering qualifications by putting these projects under the direct supervision of regional or central irrigation office staff or by implementing more stringent protocols for engineer review and approval of actions taken at the projects; and
- require, at a minimum, that irrigation project management meet twice annually with all project stakeholders—once at the end of a season and once before the next season—to provide information on project operations, including budget plans and actual annual expenditures, and to obtain feedback and input.

To obtain information on the long-term financial sustainability of each of the projects, we recommend that the Secretary direct the Assistant Secretary for Indian Affairs to conduct studies to determine both how much it would cost to financially sustain each project, and the extent to which water users on each project have the ability to pay these costs. This information will be useful to congressional decisionmakers and other interested parties in debating the long-term direction of BIA's irrigation program.

AGENCY COMMENTS AND OUR EVALUATION

We provided the Department of the Interior with a draft of this report for review and comment. However, no comments were provided in time to be included as part of this report.

Robin M. Nazzaro
Director, Natural Resources and Environment

APPENDIX I. OBJECTIVES, SCOPE, AND METHODOLOGY

We were asked to address several issues concerning the Department of the Interior's Bureau of Indian Affairs' (BIA) management of its 16 irrigation projects. Specifically, we were asked to examine (1) BIA's estimated deferred maintenance cost for its 16 irrigation projects; (2) what shortcomings, if any, exist in BIA's current management of its irrigation projects; and (3) any issues that need to be addressed to determine the long-term direction of BIA's irrigation program.

For all three objectives, we collected documentation on BIA's 16 irrigation projects from officials in each of BIA's central Irrigation, Power, and Safety of Dams offices (central irrigation offices) located in Washington, D.C., and other locations in the western United States. We also visited and collected information from each of BIA's four regional offices that oversee the 16 irrigation projects, including the Rocky Mountain, Northwest, Western, and Southwest regions. In addition, we visited 9 of the 16 projects located across all 4 regions. Specifically, we visited: (1) the Blackfeet Irrigation Project, (2) the Colorado River Irrigation Project, (3) the Crow Irrigation Project, (4) the Fort Belknap Irrigation Project, (5) the Pine River Irrigation Project, (6) the San Carlos Indian Works Irrigation Project, (7) the San Carlos Joint Works Irrigation Project, (8) the Wapato Irrigation Project, and (9) the Wind River Irrigation Project. We selected these projects based on a combination of factors aimed at maximizing our total coverage (over 50 percent of the projects), visiting at least one project in each of the regions where irrigation projects are located, visiting the project with the highest deferred maintenance cost estimate in each region using BIA's fiscal year 2004 data, and visiting what BIA considered to be the three best projects and the five worst projects. During the site visits, we collected project-specific information from BIA officials and project stakeholders including tribes and water users. We also met with and collected documentation from the Department of the Interior's Bureau of Reclamation, the federal agency charged with managing water in the western United States, for comparative purposes.

To examine BIA's estimated deferred maintenance cost for its 16 irrigation projects, we toured each of the 9 projects we visited to see examples of deferred maintenance and their impact, and we reviewed BIA's lists of deferred maintenance items and associated cost estimates for both fiscal years 2004 and 2005. We also reviewed the methodology BIA used to develop these lists and estimates and interviewed BIA staff involved in developing these lists and estimates to identify major deficiencies. Although we analyzed the cost estimates provided by BIA, we did not develop our own estimate of deferred maintenance. To assess the reliability of data we received from BIA on deferred maintenance, we interviewed officials most knowledgeable about the collection and management of these data. We reviewed the relevant controls and found them adequate. We also conducted tests of the reliability of the computerized data. On the basis of these interviews, tests, and reviews, we concluded that BIA's estimates of deferred maintenance were sufficiently reliable for the purposes of this report.

To examine what shortcomings, if any, exist in BIA's current management of its irrigation projects, we reviewed relevant federal regulations and agency guidance, and analyzed BIA-wide and project-specific management protocols and systems for the nine projects we visited. We also reviewed general guidance on internal control standards, including risk assessment, monitoring, and information and communication. We interviewed BIA officials from the central irrigation office in Washington, D.C., Colorado, Oregon, Arizona and Montana. We also interviewed BIA regional officials as well as agency and project officials associated with each of the 9 projects we visited for information on key shortcomings in BIA's management of its irrigation projects. Finally, we interviewed a variety of project stakeholders—including tribal representatives, individual Indian water users, and non-Indian water users—at each of the 9 projects we visited for information on key shortcomings in BIA's management.

Finally, to examine any issues that need to be addressed to determine the long-term direction of BIA's irrigation program, we reviewed previous studies highlighting key issues impacting the future of BIA's irrigation program. This included reviewing previous studies conducted by GAO, the Department of the Interior's Office of Inspector General, and the Bureau of Reclamation, as well as other studies conducted at the request of the Congress. We also reviewed relevant federal regulations and agency guidance, as well as historical information relevant to BIA's management of the irrigation program, including budget information and agency memos. Finally, we interviewed BIA officials from the central irrigation office, regional offices, and the 9 projects we visited for information on the key challenges impacting the long-term direction of the program. We also interviewed project stakeholders—including tribal representatives and water users—at the 9 projects we visited for information on the key issues impacting the future direction of BIA's irrigation program.

We performed our work between March 2005 and February 2006 in accordance with generally accepted government auditing standards.

APPENDIX II. PROFILES OF THE NINE IRRIGATION PROJECTS GAO VISITED

This appendix contains brief profiles of the nine irrigation projects we visited. Each project profile begins with a short overview of basic facts about the project, followed by a set of bullet points describing the key operations and maintenance concerns and the key management concerns expressed to us by BIA officials, tribal officials, or water users during our site visits.

Blackfeet Irrigation Project

The Blackfeet Irrigation Project was authorized for construction in 1907, but construction was never completed. It consists of 38,300 acres being assessed operations and maintenance fees (and 113,100 acres authorized for irrigation). The project is located in Browning, Montana on the Blackfeet Indian Reservation of Montana, home of the Blackfeet Tribe. About 60 percent of the project's land is owned by either the tribe or individual tribal members, and about 40 percent is owned by non-Indians. BIA currently estimates the project's total deferred maintenance costs to be $29,130,222. See figure 8 below for pictures of the Blackfeet Irrigation Project.

Deteriorating Canal Lining

Leaking Check Gate

Source: GAO.

Figure 8. Pictures of the Blackfeet Irrigation Project (July 2005).

Key Operations and Maintenance Concerns Expressed during Our Site Visit

- Fees are insufficient to cover the costs of project operations and maintenance.
- Weeds and overgrown vegetation are problematic and impair water flow.
- Deferring maintenance has led to bigger and more costly maintenance problems.
- Deferring maintenance decreases water efficiency and access to water.
- The project as built cannot meet the increased demand for water.

Key Management Concerns Expressed during Our Site Visit

- Communication between BIA and the water users could be improved, such as enhancing transparency, increasing involvement, and meeting separately with the tribe.
- Lack of training and expertise undermines BIA's management of the project.
- Inadequate oversight within BIA exacerbates problems associated with lack of training and expertise.
- Project staff should report to managers with expertise in irrigation and/or engineering.
- BIA protocols are too vague, such as when project staff should consult with regional or central irrigation office engineers.
- BIA needs to be able to measure water in order to better manage water deliveries and identify critical problems.
- Irrigation is a low priority for BIA.

Colorado River Irrigation Project

The Colorado River Irrigation Project was the first BIA irrigation project built, authorized for construction in 1867, but construction was never completed. It is now considered the best of BIA's 16 revenue-generating irrigation projects due, in part, to its innovative leadership and customer service attitude. The project has adopted a user fee system that measures and assesses water users based on their actual usage as well as charging water users additional fees for using more water than their individual allotment. The project is located in Parker, Arizona on the Colorado River Indian

Reservation, home of the Colorado River Indian Tribes. The project, which has a 10-month-long irrigation season, consists of 79,350 assessed acres (and 107,588 acres authorized for irrigation), and is composed entirely of Indian land—land owned by the tribe or its members. BIA currently estimates the project's total deferred maintenance costs to be $134,758,664. See figure 9 for pictures of the Colorado River Irrigation Project.

Key Operations and Maintenance Concerns Expressed during Our Site Visit

- Development leases may no longer be allowed, potentially resulting in irrigable land going un-irrigated and costing the tribe and project potential revenues.
- Replacement of deteriorating irrigation structures needed.
- Canal needs new lining due to years of deterioration and, in some cases, poor construction.
- Clearing moss and pondweed is needed lest the flow of water be impaired.
- New irrigation structures needed to regulate water flow where ditches converge.

Concrete-lined Irrigation Canal

Leaking Check Gate

Source: GAO.

Figure 9. Pictures of the Colorado River Irrigation Project (June 2005).

Key Management Concerns Expressed during Our Site Visit

- Understaffing and high turnover of project system operators adversely impact water deliveries in that there are too few system operators to deliver water in a timely manner.
- BIA procurement and contracting is time-consuming and costly.
- Annual project budget may understate actual funding because it does not include possible additional fees.
- Operations and maintenance fees can only be used to address operations and maintenance on the existing project, rather than expand the project.

Crow Irrigation Project

The Crow Irrigation Project was authorized for construction in 1890, but construction was never completed. It is one of the oldest of BIA's 16 revenue-generating irrigation projects with 38,900 acres being assessed operations and maintenance fees (and 46,460 acres authorized for irrigation). The project is located in Crow Agency, Montana on the Crow Reservation, home of the

Crow Tribe of Montana. About 56 percent of the project land is owned by either the tribe or individual tribal members, and about 44 percent is owned by individual non-Indians. BIA currently estimates the project's total deferred maintenance costs to be $54,550,496. See figure 10 for pictures of the Crow Irrigation Project.

Key Operations and Maintenance Concerns Expressed during Our Site Visit

- Fees are insufficient to cover the project's operations as well as maintenance costs.
- Weeds, overgrown vegetation, tree roots and garbage impair water flow in the canals and ditches.
- Crumbling or dilapidated irrigation structures impair water delivery.
- The repair of Rotten Grass Flume needs further work.
- Canal erosion causes sink holes and impairs water flow.
- Deferred maintenance of certain structures leads to safety concerns, such as when BIA staff must go into the canal to raise or lower broken check gates.

Abandoned Car in Deteriorated Canal

Crumbling Irrigation Structure

Source: GAO.

Figure 10. Pictures of the Crow Irrigation Project (March 2005).

Key Management Concerns Expressed during Our Site Visit

- The project's recently reassigned project manager was under-qualified, resulting in some decisions that hurt the project and undermine water delivery, such as the Rotten Grass Flume incident.
- BIA has inadequate oversight of the project manager and his decisions.
- BIA relies on "crisis-style" management rather than a long-term plan to manager project.
- Allegations that a former project manager inappropriately used fees and was not accountable for financial decisions.
- Communication breakdown between BIA and its water users.
- The project may be better managed if BIA turned over the project's management to water users or tribe.
- Irrigation is a low priority for BIA.

Fort Belknap Irrigation Project

The Fort Belknap Irrigation Project was authorized for construction in 1895, but construction was never completed. It is one of the smallest of BIA's 16 revenue-generating irrigation projects with 9,900 acres being assessed operations and maintenance fees (and 13,320 acres authorized for irrigation). The project is located in Harlem, Montana on the Fort Belknap Reservation, home of the Fort Belknap Indian Community of the Fort Belknap Reservation of Montana. About 92 percent of the land is owned by either the tribe or individual tribal members, and about 8 percent is owned by individual non-Indians. BIA currently estimates the project's total deferred maintenance costs to be $17,535,494. See figure 11 for pictures of the Fort Belknap Irrigation Project.

Key Operations and Maintenance Concerns Expressed during Our Site Visit

- Fees and appropriations are insufficient to cover the project maintenance needs.
- Weeds and overgrowth of vegetation impair water flow.
- Canal erosion caused by cattle-crossings impairs water flow.
- Deteriorated and leaking irrigation structures impair water delivery.
- Additional equipment is needed to conduct maintenance on project.
- Deferred maintenance exacerbates problems of poor farming land and low crop values.

Key Management Concerns Expressed during Our Site Visit

- Poor communication and tense relations between BIA and water users.
- Staff turnover and difficulty finding qualified staff are problematic.
- Some project staff lack adequate expertise and training to manage project.
- Lack of transparency and water management plan limits BIA accountability.
- Some water users want BIA to begin water delivery earlier in season.

Graffiti on Irrigation Structure

Overgrown Vegetation Around Dilapidated Irrigation Check Structure

Source: GAO.

Figure 11. Pictures of the Fort Belknap Irrigation Project (July 2005).

Pine River Irrigation Project

The Pine River Irrigation Project is the only one of BIA's 16 revenue-generating irrigation projects located in the Southwest region, with 11,855 acres being assessed operations and maintenance fees. Construction on the project was never completed. The project is located in Ignacio, Colorado on the Southern Ute Reservation, home to the Southern Ute Indian Tribe of the Southern Ute Reservation, Colorado. About 85 percent of the land is owned by either the tribe or individual tribal members, and about 15 percent is owned by individual non-Indians. BIA currently estimates the project's total deferred maintenance costs to be $20,133,950. See figure 12 for pictures of the Pine River Irrigation Project.

Irrigation Ditch in Need of Reshaping

Overgrown Vegetation in Irrigation Ditch.

Source: GAO.

Figure 12. Pictures of the Pine River Irrigation Project (August 2005).

Key Operations and Maintenance Concerns Expressed during Our Site Visit

- Collections from operations and maintenance fees do not provide adequate funds to properly operate and maintain the project.
- The project's operations and maintenance fees have not been raised since 1992. BIA has proposed doubling the fees from $8.50 per acre to $17.00 per acre for the 2006 irrigation season.
- The project's cash reserves were depleted in 2004.
- The project has a number of old water delivery contracts, referred to as "carriage contracts," from the 1930s that are at low fixed rates. Under some of the contracts the water users only pay $1.00 per acre to the project.

Key Management Concerns Expressed during Our Site Visit

- The practice of subsidizing the project through other BIA programs, such as Natural Resources, Roads Construction, Roads Maintenance and Realty, was scheduled to end at the end of fiscal year 2005. Alternative sources of funds must be found for the project manager and clerk positions.
- "Crisis-style" management only, no preventive maintenance.
- Project staff does not formally meet with or provide information to individual water users.
- A Bureau of Reclamation study in 1999 found that some of the water users could not afford to pay fees of $8.50 to the project and operate a profitable farming operation. BIA has not responded to the study.
- The former project manager stated that the BIA irrigation projects should be turned over to the Bureau of Reclamation.

San Carlos Indian Works Irrigation Project (Pima)

The San Carlos Indian Works Irrigation Project was authorized for construction in 1924, but construction was never completed. It is one of the newest of BIA's 16 revenue-generating irrigation projects with 50,000 acres being assessed operations and maintenance fees (and 50,546 acres authorized for irrigation). The project, also referred to as Pima, is located in Sacaton, Arizona on the Gila River Indian Reservation, home of the Gila River Indian Community. It is served both by its own infrastructure and by that of the San Carlos Joint Works Irrigation Project. The project land is generally owned by the tribe or tribal members, with about 99 percent of the land owned by either the tribe or individual tribal members, and about 1 percent owned by individual non-Indians. BIA currently estimates Pima's total deferred maintenance costs to be $62,865,503. See figure 13 for pictures of the San Carlos Indian Works Irrigation Project.

Key Operations and Maintenance Concerns Expressed during Our Site Visit

- Inefficiency in water delivery results in fewer water users being able to receive water, leading to idle acreage in some cases.

- Clearing tumbleweeds and other vegetation that can clog culverts are a recurring problem and represents a large part of the project's spending on operations and maintenance.
- Erosion is a continuing problem, in part, because the canal is used for both water deliveries as well as drainage.
- BIA staff has a "wish list" of items that would bring the project into top condition, extending beyond the basic deferred maintenance.
- Project infrastructure may not have the capacity to deliver water to all potential water users.

Concrete-lined Irrigation Canal

Canal with Vegetation Growth

Source: GAO.

Figure 13. Pictures of the San Carlos Indian Works Irrigation Project (June 2005).

Key Management Concerns Expressed during Our Site Visit

- 2007 turnover to water users is still underway.
- Insufficient reserve funds means that project staff may not have enough money to conduct needed maintenance towards the end of the year.
- Vacancies are a constant problem at the project, leaving too few staff to conduct project maintenance.
- BIA is too slow to respond to water users' requests for repairs.

San Carlos Joint Works Irrigation Project (Coolidge)

The San Carlos Joint Works Irrigation Project was authorized for construction in 1924, but construction was never completed. It provides water to non-Indian irrigators as well as the San Carlos Indian Works Irrigation Project. It consists of 100,000 acres being assessed operations and maintenance fees (and 100,546 acres authorized for irrigation), with 50 percent of the land owned by non-Indian irrigators and 50 percent owned by Indian irrigators (in the form of the San Carlos Indian Works Irrigation Project). The project is located in Coolidge, Arizona. BIA currently estimates Coolidge's total deferred maintenance costs to be $5,775,427. See figure 14 for pictures of the San Carlos Joint Works Irrigation Project.

Deteriorating China Wash Flume

Concrete that Fell from Flume

Weeds in Irrigation Canal

Source: GAO.

Figure 14. Pictures of the San Carlos Joint Works Irrigation Project (June 2005).

Key Operations and Maintenance Concerns Expressed during Our Site Visit

- Lack of certainty in BIA's ability to deliver requested water to all water users has led some to purchase additional water from outside of the project.
- Silt removal from irrigation canals and ditches is a recurring problem, leading BIA to purposefully over-excavate the main canal each year in an attempt to catch excess silt that can clog culverts and prevent water delivery impairments.
- Repair of China Wash Flume is an expensive undertaking, but the flume's failure could jeopardize water deliveries for much of the project.
- Removal of weeds to prevent clogged culverts is a recurring problem for the project.

Key Management Concerns Expressed during Our Site Visit

- 2007 turnover to water users is under way but not finalized.
- Lawsuit against BIA's increase in operations and maintenance fees resulted in some water delivery delays while the lawsuit is pending.
- Contracting delays within BIA have resulted in postponed project maintenance.
- Turnover of BIA staff and lack of water user inclusion in project decisionmaking impedes effective communication.
- BIA lacks accountability to water users in terms of how it spends operations and maintenance fees.

Wapato Irrigation Project

The Wapato Irrigation Project is one of the oldest and largest of BIA's 16 revenue-generating irrigation projects with 96,443 acres being assessed operations and maintenance fees (and 145,000 acres authorized for irrigation). It was authorized for construction in 1904, but construction was never completed. The project is located in Yakima, Washington on the Yakama Reservation, home of the Confederated Tribes and Bands of the Yakama Nation. About 60 percent of the project land is owned by either the tribe or individual tribal members, and about 40 percent is owned by individual non-

Indians. BIA currently estimates the project's total deferred maintenance costs to be $183,128,886. See figure 15 for pictures of the Wapato Irrigation Project.

Weed Clearing Machine

Irrigation Canal with Crumbling Concrete Lining

Source: GAO.

Figure 15. Pictures of the Wapato Irrigation Project (April 2005).

Key Operations and Maintenance Concerns Expressed during Our Site Visit

- Deterioration of project prevents some water users from receiving water.
- Lack of regular project maintenance has led many water users to make repairs on their own in order to irrigate crops.
- Water users claim that project staff performs inadequate or faulty repairs, resulting in wasted operations and maintenance payments or the need for water users to fix the sloppy repairs.
- Fees are insufficient because (a) rates have been set too low, and (b) the tribe's appeal of BIA's operations and maintenance bills since 2001 has decreased income by at least $2 million annually because the agency will not collect on these bills or issue subsequent bills until the matters raised in the appeal are resolved.
- Fees are insufficient to cover both maintenance and administrative costs, such as salaries and benefits, leading to suggestions that BIA cover such costs.

Key Management Concerns Expressed during Our Site Visit

- Understaffing due to inadequate funds and difficulty in finding qualified staff has resulted in too few staff to operate and maintain project.
- BIA relies on "crisis-style" management to manage project, resulting in a lack of planning and preventive maintenance.
- Water users lack voice in project decisionmaking, resulting in concerns about limited accountability of project staff to its water users.
- Alleged errors with operations and maintenance billing—such as BIA billing dead landowners and BIA overbilling living landowners—led the tribe and its members to appeal BIA's billing of operations and maintenance fees. Resolution of these appeals is still pending within the agency. BIA will not collect on these bills or issue subsequent bills until the matters raised in the appeal are resolved.

Wind River Irrigation Project

The Wind River Irrigation Project was authorized for construction in 1905, but construction was never completed. It is one of BIA's 16 revenue-generating irrigation projects with 38,300 acres being assessed operations and maintenance fees (and 51,000 acres authorized for irrigation). The project is located in Fort Washakie, Wyoming on the Wind River Reservation, home of the Arapaho Tribe of the Wind River Reservation and the Shoshone Tribe of the Wind River Reservation. About 67 percent of the project land is owned by either the tribe or individual tribal members, and about 33 percent is owned by individual non-Indians. BIA currently estimates the project's total deferred maintenance costs to be $84,956,546. See figure 16 for pictures of the Wind River Irrigation Project.

Beaver Holes and Overgrown Vegetation in Canal

Cattle- Crossing Eroding Irrigation Canal

Source: GAO.

Figure 16. Pictures of the Wind River Irrigation Project (July 2005).

Key Operations and Maintenance Concerns Expressed during Our Site Visit

- Weeds and tree roots impair water flow and lead to seepage.
- Cattle-crossings erode canal banks and impair water flow.
- Deteriorating irrigation infrastructure impairs water delivery.
- Additional water storage and improved efficiency needed to meet demand for water.
- Deferring maintenance undermines long-term sustainability of project.
- BIA financial management may limit ability of project staff to conduct needed maintenance in short maintenance season.

Key Management Concerns Expressed during Our Site Visit

- BIA relies on "crisis-style" management and "band-aid" solutions rather than a long-term plan to manage project.
- Poor communication between BIA and water users.
- Water users are not involved enough in project decisionmaking.
- Supervision of project staff is insufficient and BIA is not accountable to water users.
- Turnover of BIA staff is problematic.
- Some water users want to manage all or part of the project.

End Notes

[1] U.S. Department of the Interior, Office of the Inspector General, *Indian Irrigation Projects, Bureau of Indian Affairs*, 96-I-641, (Washington D.C.: March 1996); U.S. Department of the Interior, Office of the Inspector General, *Operations and Maintenance Assessments of Indian Irrigation Projects, Bureau of Indian Affairs,* W-IA-BIA-12-86, (Washington D.C.: Feb. 1988).

[2] 25 C.F.R. § 171.1(c).

[3] 25 C.F.R. § 171.1(d).

[4] Act of July 4, 1884, 23 Stat. 76, 94 (1884).

[5] See 70 *Fed. Reg.* 57889 (Oct. 4, 2005) for the 2005 operations and maintenance fees for the projects as well as the proposed fees for 2006.

[6] 25 C.F.R. § 171.1 (c), (d).

[7] GAO, *Indian Programs: BIA's Management of the Wapato Irrigation Project*, GAO/RCED97-124 (Washington D.C.: May 28, 1997); U.S. Department of the Interior, Office of the Inspector General, *Indian Irrigation Projects, Bureau of Indian Affairs*, 96-I-641, (Washington D.C.: March 1996).

[8] BIA defines deferred maintenance as "maintenance that was not performed when it should have been or when it was scheduled and which, therefore, was put off or delayed for a future period (adapted from [Federal Accounting Standards Advisory Board Bulletin] No. 6). This

includes facility deficiencies where there is non-compliance to codes (e.g., life safety, [Americans with Disabilities Act, Occupational Safety and Health Administration], environmental, etc.) and other regulatory or Executive Order compliance requirements."

[9] BIA implemented MAXIMO agencywide, not just for its irrigation projects, to help identify and track deferred maintenance.

[10] This condition assessment, entitled *Final Report: Engineering Evaluation of Existing Conditions, Flathead Agency Irrigation Division (FAID)* was prepared by HKM Engineering for the Confederated Salish and Kootenai Tribes. BIA did not fund this condition assessment.

[11] GAO, Internal Control Standards: Internal Control Management and Evaluation Tool, GAO-01-1008G (Washington, D.C.: Aug. 2001).

[12] Survey of Conditions of the Indians in the United States: Hearings Before a Subcommittee of the Senate Committee on Indian Affairs, 71st Cong., 2d Sess. at 2219-20 (1930).

[13] 70 *Fed. Reg.* 57889, 57893 (Oct. 4, 2005).

[14] These funds were specifically appropriated for five irrigation projects—Crow, Fort Peck, Fort Belknap, Blackfeet and Wind River—and one irrigation system.

End Note for Appendix II

[1] GAO referred these allegations to the Department of the Interior's Office of the Inspector General in August 2005.

INDEX

A

B

C

D

E

F

G

H

I

L

M

N

O

P

Q

R

S

T

U

V

W